"十三五"应用型人才培养规划教材

# Struts 2 框架实用教程

陈恒 主编　徐琳宏 陶永鹏 张术梅 副主编

清华大学出版社
北京

## 内容简介

本书是一本应用教程,以大量的实例介绍了 Struts 2 框架应用的基本思想、方法和技术,同时配备了相应的实践环节巩固 Struts 2 应用开发的方法和技术,力图达到"做中学,学中做"。

全书共分 10 章,内容包括 Struts 2 入门、Struts 2 的 Action、Struts 2 的类型转换、Struts 2 的拦截器、表达式语言 OGNL、Struts 2 的标签、Struts 2 的输入校验、Struts 2 的国际化、文件的上传和下载以及名片管理系统的设计与实现。书中实例侧重实用性和启发性,趣味性强、分布合理、通俗易懂,使读者能够快速掌握 Struts 2 框架应用的基础知识、编程技巧以及完整的开发体系,为适应实战应用打下坚实的基础。

本书可以作为大学计算机及相关专业的教材或教学参考书,也可以作为 Struts 2 应用开发人员的参考用书。

本书封面贴有清华大学出版社防伪标签,无标签者不得销售。
版权所有,侵权必究。侵权举报电话:010-62782989　13701121933

图书在版编目(CIP)数据

Struts 2 框架实用教程/陈恒主编.—北京:清华大学出版社,2017
("十三五"应用型人才培养规划教材)
ISBN 978-7-302-47602-3

Ⅰ.①S… Ⅱ.①陈… Ⅲ.①软件工具-程序设计-高等学校-教材 Ⅳ.①TP311.56

中国版本图书馆 CIP 数据核字(2017)第 153463 号

责任编辑:田在儒
封面设计:王跃宇
责任校对:赵琳爽
责任印制:李红英

出版发行:清华大学出版社
网　　址:http://www.tup.com.cn, http://www.wqbook.com
地　　址:北京清华大学学研大厦 A 座　　　　邮　编:100084
社 总 机:010-62770175　　　　　　　　　　邮　购:010-62786544
投稿与读者服务:010-62776969, c-service@tup.tsinghua.edu.cn
质量反馈:010-62772015, zhiliang@tup.tsinghua.edu.cn
课件下载:http://www.tup.com.cn,010-62770175-4278

印 装 者:北京鑫海金澳胶印有限公司
经　　销:全国新华书店
开　　本:185mm×260mm　　印　张:15.75　　字　数:357 千字
版　　次:2017 年 8 月第 1 版　　　　　　　印　次:2017 年 8 月第 1 次印刷
印　　数:1~3000
定　　价:39.00 元

产品编号:073628-01

尽管已经有许多与 Struts 2 框架有关的书籍,但大部分书籍过于注重知识的系统性,使得知识体系结构过于全面、庞大。这种知识体系结构过于庞大的书籍不太适合作为高校计算机相关专业的教材。同时,许多教师在教学过程中,非常希望教材本身能引导学生尽可能多地参与到教学活动中,因此本书的重点不是简单地介绍 Struts 2 框架的基础知识,而是加入了大量的实例与实践环节。读者通过本书可以快速地掌握 Struts 2 框架技术,提高开发 Struts 2 应用的能力。全书共 10 章,其各章的具体内容如下。

第 1 章概括地介绍了 MVC 的设计思想,详细地讲解了 Struts 2 应用开发框架的构建。

第 2 章讲解了 Action 的编写方式、接收用户数据的方式、在 Action 中如何访问 Servlet API 以及 Action 中常见的结果类型,是本书的重点内容之一。

第 3 章讲解了 Struts 2 的类型转换,包括内置的类型转换器和自定义类型转换器。

第 4 章讲解了 Struts 2 的拦截器,包括内置的拦截器和自定义拦截器。

第 5 章讲解了表达式语言 OGNL,包括 OGNL 语法基础、值栈的概念以及 OGNL 表达式的应用。

第 6 章详细地讲解了 Struts 2 标签,包括非 UI 标签和 UI 标签,是本书的重点内容之一。

第 7 章讲解了 Struts 2 框架的输入校验体系,包括手动编程校验和校验框架校验。

第 8 章讲解了 Struts 2 国际化的实现方法,包括 JSP 页面国际化、校验信息国际化以及 Action 信息国际化等内容。

第 9 章详细地讲解了如何使用 Struts 2 框架进行文件的上传与下载,包括单文件上传、多文件上传以及文件下载。

第 10 章是本书的重点内容之一,将前面章节的知识进行一个大综合,详细地讲解了如何使用 Struts 2 框架来开发一个 Web 应用(名片管理系统)的过程。

本书特别注重引导学生参与课堂教学活动,适合作为大学计算机及相关专业的教材或教学参考书,也适合作为 Struts 2 应用开发人员的参考用书。

为了便于教学，本书配有教学课件、源代码以及实践环节与课后习题参考答案，读者可从清华大学出版社网站免费下载。

由于编者水平有限，书中难免会有不足之处，敬请广大读者批评指正。

编 者

2017 年 2 月

# 第 1 章　Struts 2 入门 ……………………………………………………………………… 1

## 1.1　Struts 2 的工作环境 …………………………………………………………… 1
### 1.1.1　核心知识 ……………………………………………………………… 1
### 1.1.2　能力目标 ……………………………………………………………… 4
### 1.1.3　任务驱动 ……………………………………………………………… 4
### 1.1.4　实践环节 ……………………………………………………………… 11
## 1.2　第一个 Struts 2 应用 …………………………………………………………… 12
### 1.2.1　核心知识 ……………………………………………………………… 12
### 1.2.2　能力目标 ……………………………………………………………… 12
### 1.2.3　任务驱动 ……………………………………………………………… 13
### 1.2.4　实践环节 ……………………………………………………………… 18
## 1.3　本章小结 ………………………………………………………………………… 19
习题 1 ……………………………………………………………………………………… 19

# 第 2 章　Struts 2 的 Action ……………………………………………………………… 20

## 2.1　Action 的创建与配置 …………………………………………………………… 20
### 2.1.1　核心知识 ……………………………………………………………… 20
### 2.1.2　能力目标 ……………………………………………………………… 24
### 2.1.3　任务驱动 ……………………………………………………………… 24
### 2.1.4　实践环节 ……………………………………………………………… 26
## 2.2　Action 接收请求参数 …………………………………………………………… 26
### 2.2.1　核心知识 ……………………………………………………………… 27
### 2.2.2　能力目标 ……………………………………………………………… 29
### 2.2.3　任务驱动 ……………………………………………………………… 30
### 2.2.4　实践环节 ……………………………………………………………… 31
## 2.3　Action 访问 Servlet API ………………………………………………………… 32
### 2.3.1　核心知识 ……………………………………………………………… 32
### 2.3.2　能力目标 ……………………………………………………………… 37
### 2.3.3　任务驱动 ……………………………………………………………… 37

    2.3.4 实践环节 ………………………………………………………… 38
  2.4 Action 中常见的结果类型 ………………………………………………… 39
    2.4.1 核心知识 ………………………………………………………… 39
    2.4.2 能力目标 ………………………………………………………… 42
    2.4.3 任务驱动 ………………………………………………………… 42
    2.4.4 实践环节 ………………………………………………………… 44
  2.5 本章小结 ………………………………………………………………… 46
  习题 2 ………………………………………………………………………… 46

## 第 3 章 Struts 2 的类型转换 ……………………………………………………… 48

  3.1 Struts 2 内置的类型转换器 ……………………………………………… 48
    3.1.1 核心知识 ………………………………………………………… 48
    3.1.2 能力目标 ………………………………………………………… 52
    3.1.3 任务驱动 ………………………………………………………… 52
    3.1.4 实践环节 ………………………………………………………… 53
  3.2 自定义类型转换器 ……………………………………………………… 53
    3.2.1 核心知识 ………………………………………………………… 56
    3.2.2 能力目标 ………………………………………………………… 59
    3.2.3 任务驱动 ………………………………………………………… 59
    3.2.4 实践环节 ………………………………………………………… 66
  3.3 本章小结 ………………………………………………………………… 66
  习题 3 ………………………………………………………………………… 66

## 第 4 章 Struts 2 的拦截器 ……………………………………………………… 67

  4.1 拦截器的定义与配置 …………………………………………………… 67
    4.1.1 核心知识 ………………………………………………………… 67
    4.1.2 能力目标 ………………………………………………………… 75
    4.1.3 任务驱动 ………………………………………………………… 75
    4.1.4 实践环节 ………………………………………………………… 78
  4.2 使用自定义拦截器完成权限验证 ……………………………………… 80
  4.3 本章小结 ………………………………………………………………… 83
  习题 4 ………………………………………………………………………… 83

## 第 5 章 表达式语言 OGNL ……………………………………………………… 84

  5.1 OGNL 基础 ……………………………………………………………… 84
    5.1.1 核心知识 ………………………………………………………… 86
    5.1.2 能力目标 ………………………………………………………… 92
    5.1.3 任务驱动 ………………………………………………………… 92
    5.1.4 实践环节 ………………………………………………………… 93

5.2　OGNL 基本语法 …………………………………………………………… 94
　　　　5.2.1　核心知识 …………………………………………………………… 94
　　　　5.2.2　能力目标 …………………………………………………………… 99
　　　　5.2.3　任务驱动 …………………………………………………………… 99
　　　　5.2.4　实践环节 …………………………………………………………… 101
　　5.3　本章小结 ………………………………………………………………… 105
　习题 5 …………………………………………………………………………… 105

第 6 章　Struts 2 的标签 ……………………………………………………………… 107
　　6.1　数据标签 ………………………………………………………………… 107
　　　　6.1.1　核心知识 …………………………………………………………… 108
　　　　6.1.2　能力目标 …………………………………………………………… 118
　　　　6.1.3　任务驱动 …………………………………………………………… 118
　　　　6.1.4　实践环节 …………………………………………………………… 119
　　6.2　流程控制标签 …………………………………………………………… 120
　　　　6.2.1　核心知识 …………………………………………………………… 120
　　　　6.2.2　能力目标 …………………………………………………………… 128
　　　　6.2.3　任务驱动 …………………………………………………………… 128
　　　　6.2.4　实践环节 …………………………………………………………… 131
　　6.3　UI 标签 ………………………………………………………………… 131
　　　　6.3.1　核心知识 …………………………………………………………… 132
　　　　6.3.2　能力目标 …………………………………………………………… 140
　　　　6.3.3　任务驱动 …………………………………………………………… 140
　　　　6.3.4　实践环节 …………………………………………………………… 144
　　6.4　本章小结 ………………………………………………………………… 144
　习题 6 …………………………………………………………………………… 144

第 7 章　Struts 2 的输入校验 ………………………………………………………… 146
　　7.1　手动编程校验 …………………………………………………………… 146
　　　　7.1.1　核心知识 …………………………………………………………… 146
　　　　7.1.2　能力目标 …………………………………………………………… 150
　　　　7.1.3　任务驱动 …………………………………………………………… 150
　　　　7.1.4　实践环节 …………………………………………………………… 152
　　7.2　校验框架校验 …………………………………………………………… 153
　　　　7.2.1　核心知识 …………………………………………………………… 153
　　　　7.2.2　能力目标 …………………………………………………………… 157
　　　　7.2.3　任务驱动 …………………………………………………………… 157
　　　　7.2.4　实践环节 …………………………………………………………… 161
　　7.3　本章小结 ………………………………………………………………… 161

习题 7 ······ 161

## 第 8 章 Struts 2 的国际化 ······ 162

### 8.1 程序国际化概述 ······ 162
8.1.1 核心知识 ······ 162
8.1.2 能力目标 ······ 165
8.1.3 任务驱动 ······ 165
8.1.4 实践环节 ······ 167

### 8.2 Struts 2 的国际化方法 ······ 167
8.2.1 核心知识 ······ 167
8.2.2 能力目标 ······ 171
8.2.3 任务驱动 ······ 171
8.2.4 实践环节 ······ 173

### 8.3 用户自定义切换语言示例 ······ 173
### 8.4 本章小结 ······ 177
习题 8 ······ 177

## 第 9 章 文件的上传和下载 ······ 179

### 9.1 Struts 2 文件上传 ······ 179
9.1.1 核心知识 ······ 179
9.1.2 能力目标 ······ 185
9.1.3 任务驱动 ······ 185
9.1.4 实践环节 ······ 189

### 9.2 Struts 2 文件下载 ······ 190
### 9.3 本章小结 ······ 193
习题 9 ······ 194

## 第 10 章 名片管理系统的设计与实现 ······ 195

### 10.1 系统设计 ······ 195
10.1.1 系统功能需求 ······ 195
10.1.2 系统模块划分 ······ 195

### 10.2 数据库设计 ······ 196
10.2.1 数据库概念结构设计 ······ 196
10.2.2 数据库逻辑结构设计 ······ 196

### 10.3 系统管理 ······ 197
10.3.1 导入相关的 jar 包 ······ 197
10.3.2 JSP 页面管理 ······ 197
10.3.3 包管理 ······ 201
10.3.4 配置文件管理 ······ 201

## 10.4 组件设计 ··· 203
### 10.4.1 工具类 ··· 203
### 10.4.2 拦截器类 ··· 204
### 10.4.3 数据库操作 ··· 204
## 10.5 名片管理 ··· 211
### 10.5.1 Action 的实现 ··· 211
### 10.5.2 添加名片 ··· 216
### 10.5.3 查询名片 ··· 218
### 10.5.4 修改名片 ··· 221
### 10.5.5 删除名片 ··· 225
## 10.6 用户相关 ··· 228
### 10.6.1 Action 的实现 ··· 228
### 10.6.2 注册 ··· 230
### 10.6.3 登录 ··· 233
### 10.6.4 修改密码 ··· 235
### 10.6.5 基本信息 ··· 237
## 10.7 安全退出 ··· 238
## 10.8 本章小结 ··· 238

# 参考文献 ··· 239

# Struts 2 入门

**主要内容**

(1) Struts 2 的体系结构。
(2) Struts 2 应用开发环境的构建。
(3) 第一个 Struts 2 应用。

MVC 思想将一个应用分成三个基本部分：Model(模型)、View(视图)和 Controller(控制器)，让这三个部分以最低的耦合进行协同工作，从而提高应用的可扩展性及可维护性。

Struts 2 是一款优秀的基于 MVC 思想的应用框架，是 Apache Struts 和 WebWork 组合产生的新产品，是最灵活、最简单的 MVC 组件。

## 1.1 Struts 2 的工作环境

 核心知识

所谓"工欲善其事，必先利其器"，在使用 Struts 2 框架进行 Web 开发前，需要构建其开发环境。安装 Struts 2 之前，需要事先安装 JDK 和 Web 服务器。

### 1. JDK

构建 Struts 2 的开发环境，首先安装并配置 JDK(本书采用的 JDK 是 jdk-7u3-windows-x64.exe)，至于 JDK 的安装和环境变量的配置，本书就不再讲解。

### 2. Web 服务器

目前，比较常用的 Web 服务器包括 Tomcat、JRun、Resin、WebSphere、WebLogic 等，本书采用的是 Tomcat 7。

登录 Apache 软件基金会的官方网站 http://jakarta.apache.org/tomcat，下载 Tomcat 7 的免安装版(apache-tomcat-7.0.57.zip)。登录网站后，首先在 Download 里选择 Tomcat 7，然后在 Binary Distributions 的 Core 中选择 zip 即可。

安装 Tomcat 之前需要事先安装 JDK 并配置系统环境变量 Java_Home。将下载的 apache-tomcat-7.0.57.zip 解压到磁盘的某个分区中，比如解压到 D:\，解压缩后将出现如图 1.1 所示的目录结构。

图 1.1 Tomcat 目录结构

执行 Tomcat 根目录中 bin 文件夹中的 startup.bat 来启动 Tomcat 服务器。执行 startup.bat 启动 Tomcat 服务器会占用一个 MS-DOS 窗口,出现如图 1.2 所示的界面,如果关闭当前 MS-DOS 窗口将关闭 Tomcat 服务器。

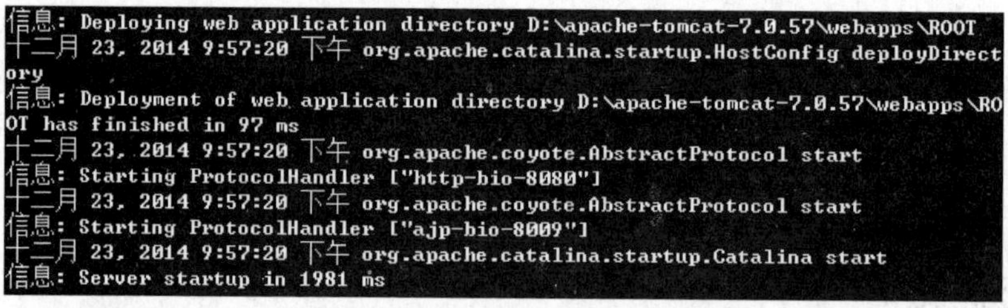

图 1.2 执行 startup.bat 启动 Tomcat 服务器时的 MS-DOS 窗口

Tomcat 服务器启动后,在浏览器的地址栏中输入:http://localhost:8080,将出现如图 1.3 所示的 Tomcat 测试页面。

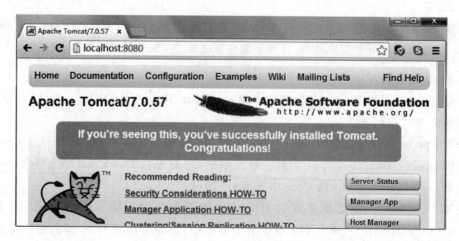

图 1.3 Tomcat 测试页面

### 3. Struts 2 的下载与安装

在浏览器中打开网址：http://struts.apache.org/download.cgi，下载 Struts 2（本书采用的是 2.3.24 版本），如图 1.4 所示。

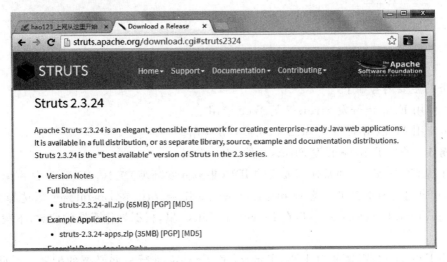

图 1.4　Struts 2 下载页面

通常建议读者下载 Struts 2 的完整版（Full Distribution），将下载到的 zip 文件解压缩，解压缩后的目录结构如图 1.5 所示。

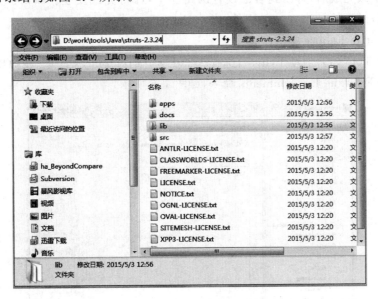

图 1.5　Struts 2 目录结构

（1）apps：该文件夹下包含了基于 Struts 2 的示例应用，这些示例应用对于读者来说是非常有用的资料。

（2）docs：该文件夹下包含了 Struts 2 的相关文档，包括 Struts 2 的快速入门、Struts 2 的文档，以及 API 文件等内容。

（3）lib：该文件夹下包含了 Struts 2 框架的核心类库，以及 Struts 2 的第三方插件类库。

（4）src：该文件夹下包含了 Struts 2 框架的全部源代码。

### 1.1.2 能力目标

安装与配置 Struts 2 的开发与运行环境。

### 1.1.3 任务驱动

任务的主要内容如下。

（1）使用 Eclipse 开发 Struts 2 的 Web 应用。

（2）使用 MyEclipse 开发 Struts 2 的 Web 应用。

**任务 1**　使用 Eclipse 开发 Struts 2 的 Web 应用

为了提高开发效率，通常还需要安装 IDE（集成开发环境）工具。Eclipse 是一个可用于开发 Web 应用的 IDE 工具。登录 http://www.eclipse.org，在 Downloads 里选择 Eclipse IDE for Java EE Developers，然后在 Download Links 里，根据操作系统的位数，下载相应的 Eclipse。

使用 Eclipse 之前，需要对 JDK、Tomcat 和 Eclipse 进行一些必要的配置。因此，在安装 Eclipse 之前，应该事先安装 JDK 和 Tomcat（见 1.1.1 小节）。

1）安装 Eclipse

Eclipse 下载完成后，解压到自己设置的路径下，即可完成安装。Eclipse 安装后，双击 Eclipse 安装目录下的 eclipse.exe 文件，启动 Eclipse。

2）Tomcat 在 Eclipse 中的配置

（1）配置 Tomcat。启动 Eclipse，选择 Window→Preferences 命令，在弹出的对话框中选择 Server→Runtime Environments 命令，如图 1.6 所示。

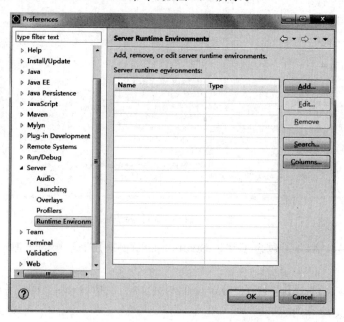

图 1.6　Tomcat 配置界面

(2) 单击 Add 按钮后,弹出如图 1.7 所示的 New Server Runtime Environment 窗口,在此可以配置各种版本的 Web 服务器。

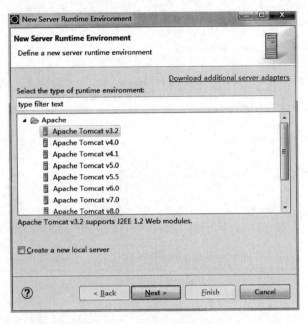

图 1.7　New Server Runtime Environment 窗口

(3) 选择 Apache Tomcat v7.0 服务器版本,单击 Next 按钮,进入如图 1.8 所示窗口。

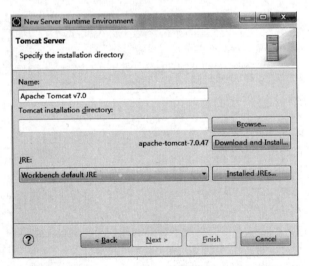

图 1.8　选择 Tomcat 目录

(4) 单击 Browse 按钮,选择 Tomcat 的目录,单击 Finish 按钮即可完成 Tomcat 配置。

3) 使用 Eclipse 开发 Struts 2 的 Web 应用

使用 Eclipse 开发一个 Struts 2 的 Web 应用,需要如下 3 个步骤。

(1) 创建 Web 应用。

① 启动 Eclipse,进入 Eclipse 开发界面。

② 选择主菜单中的 File→New→Project 命令,打开 New Project 对话框,在该对话框中选择 Web 节点下的 Dynamic Web Project 子节点,如图 1.9 所示。

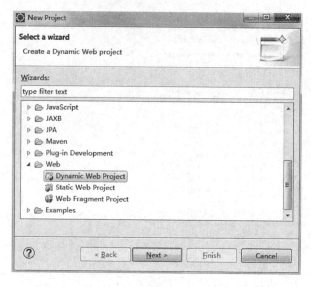

图 1.9　New Project 窗口

③ 单击 Next 按钮,打开 New Dynamic Web Project 窗口,在该对话框的 Project name 文本框中输入项目名称,这里为 firstProject。选择 Target runtime 区域中的服务器,如图 1.10 所示。

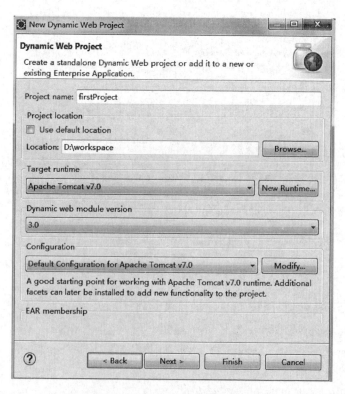

图 1.10　New Dynamic Web Project 窗口

④ 两次单击 Next 按钮后,选中 Generate web.xml deployment descriptor 复选框,如图 1.11 所示。

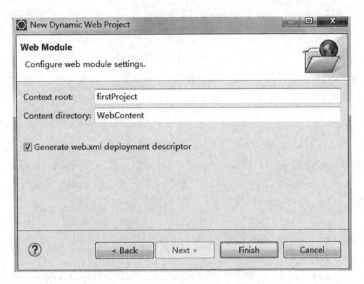

图 1.11　选择生成 web.xml

⑤ 单击 Finish 按钮,完成项目 firstProject 的创建。此时在 Eclipse 平台左侧,将显示项目 firstProject,依次展开各节点,可显示如图 1.12 所示的目录结构。

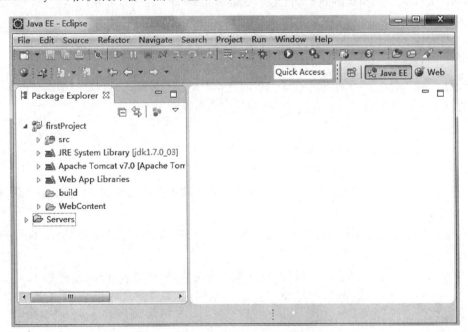

图 1.12　项目 firstProject 的目录结构

(2) 为 Web 应用添加 Struts 2 的类库。

将 Struts 2 解压缩后的 lib 目录下的 commons-fileupload-1.3.1.jar、commons-io-2.2.jar、commons-lang3-3.2.jar、freemarker-2.3.22.jar、javassist-3.11.0.GA.jar、ognl-3.0.6.jar、 struts2-core-2.3.24.jar、xwork-core-2.3.24.jar 等必需类库复制到 Web 应用的 WEB-INF/lib 目录下，也可以从 Struts 2 解压缩后的 apps 目录下的 struts2-blank 的 lib 中复制 Struts 2 类库。如果需要使用 Struts 2 的更多特性，则将相应的 JAR 文件复制到 Web 应用的 WEB-INF/lib 目录下即可。

(3) 在 web.xml 文件中配置 Struts 2 的核心过滤器。

要想在开发 Web 应用时使用 Struts 2 框架，还需要在 web.xml 中配置 Struts 2 的核心过滤器，代码如下所示：

```xml
<?xml version="1.0" encoding="UTF-8"?>
<web-app xmlns:xsi="http://www.w3.org/2001/XMLSchema-instance" xmlns="http://java.sun.com/xml/ns/javaee" xsi:schemaLocation="http://java.sun.com/xml/ns/javaee http://java.sun.com/xml/ns/javaee/web-app_3_0.xsd" id="WebApp_ID" version="3.0">
  <!-- 配置 Struts 2 的核心过滤器 -->
  <filter>
    <filter-name>Struts2</filter-name>
    <filter-class>
        org.apache.struts2.dispatcher.ng.filter.StrutsPrepareAndExecuteFilter
    </filter-class>
  </filter>
  <filter-mapping>
    <filter-name>Struts2</filter-name>
    <url-pattern>/*</url-pattern>
  </filter-mapping>
</web-app>
```

经过上面 3 个步骤，已经可以在 firstProjectWeb 应用中使用 Struts 2 的基本功能了。

**任务 2** 使用 MyEclipse 开发 Struts 2 的 Web 应用

1) 安装 MyEclipse

MyEclipse 是一款商业的、基于 Eclipse 的 Java EE 集成开发工具，不是免费产品，官方中文站点是 http://www.myeclipsecn.com/。

本书选择下载安装 MyEclipse 2014 版本。采用默认安装即可，具体过程不再赘述。

2) 启动 MyEclipse

在运行菜单中选择运行命令，即可启动 MyEclipse，启动过程与 Eclipse 类似，不再介绍。

3) 注册 MyEclipse

因为 MyEclipse 是商业软件，所以完成软件安装后，需要进行注册，步骤如下。

(1) 在 MyEclipse 菜单中选择 Window→Preferences 命令，弹出 Preferences 窗口，如图 1.13 所示。

(2) 选择 MyEclipse→Subscription 命令，进入 MyEclipse 注册页面，如图 1.14 所示。

(3) 单击 Enter Subscription 按钮，输入注册信息，即可完成软件的注册。

4) Tomcat 在 MyEclipse 中的配置

从 Windows 菜单中选择 Preferences 命令，弹出 Preferences 窗口，在该窗口中选择 MyEclipse→Servers→Tomcat→Tomcat 7.x，打开如图 1.15 所示的窗口。

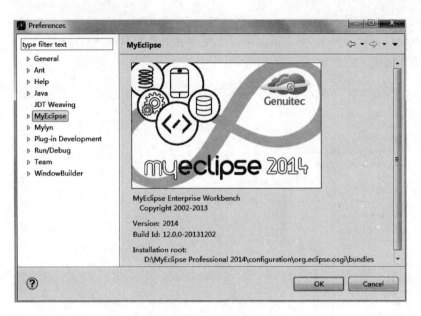

图 1.13　MyEclipse 属性窗口

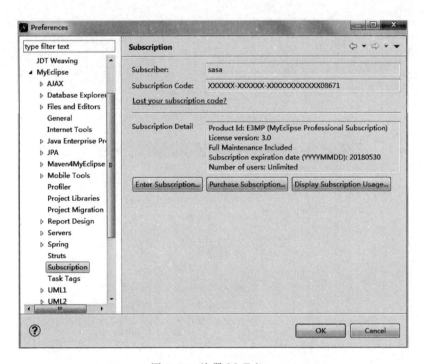

图 1.14　注册 MyEclipse

在图 1.15 中，选中 Tomcat 7.x server 区域的 Enable 单选按钮，并在 Tomcat home directory 的设置中单击 Browse 按钮，选择 Tomcat 的安装路径，单击 Apply 按钮后，再单击 OK 按钮即可完成 Tomcat 的配置。

5) 创建 Web 应用

在 MyEclipse 窗口左边的 Package Explorer 空白处，右击，在弹出的快捷菜单中选择

New Web Project 命令，进入如图 1.16 所示的窗口，在 Project name 的文本框中输入工程的名称"firstProject"，在 Target runtime 的下拉列表中选择 Apache Tomcat v7.0，最后单击 Finish 按钮即完成工程的创建。

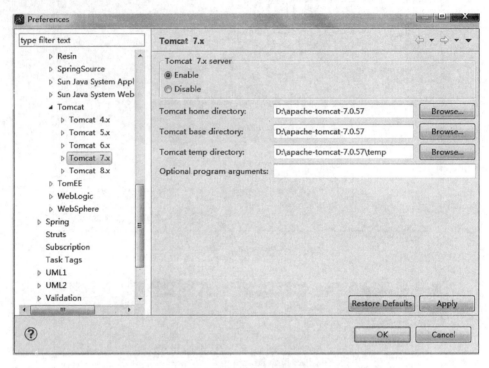

图 1.15　配置 Tomcat

图 1.16　创建 Web 应用

6）为 Web 应用添加 Struts 2 的支持

选中 Web 应用的项目名称 firstProject，右击，在弹出的快捷菜单中选择 MyEclipse→Project Facets［Capabilities］→Install Apache Struts(2.x) Facet 命令，进入如图 1.17 所示的窗口。

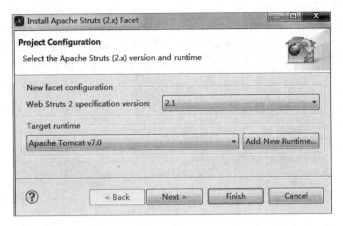

图 1.17　Web 应用配置

在图 1.17 中单击 Next 按钮，进入如图 1.18 所示的窗口，在该窗口的 URL pattern 中，选中 *.action 单选按钮，单击 Finish 按钮完成添加 Struts 2 的支持。

图 1.18　添加 Struts 2 的支持

通过 MyEclipse 给 Web 应用 firstProject 添加 Struts 2 的支持后，MyEclipse 会在工程 firstProject 的 web.xml 文件中自动配置 Struts 2 的核心过滤器，并在 src 目录下生成 struts.xml 文件，同时自动将 Struts 2 类库添加到 Web 工程中的 lib 下。不过，要注意的是这里添加的 Struts 2 类库是 MyEclipse 自带的 Struts 2 类库。当然了，也可以手动复制 1.1.1 节中的 Struts 2 类库，具体方法与任务 1 中相同，这里不再说明。

### 1.1.4　实践环节

（1）下载 Struts 2 最新版本。

（2）在 Eclipse 中为 Web 应用搭建 Struts 2 开发环境。

（3）在 MyEclipse 中为 Web 应用搭建 Struts 2 开发环境。

## 1.2 第一个 Struts 2 应用

### 1.2.1 核心知识

Struts 2 体系与 Struts 1 体系的差别非常大,因为 Struts 2 使用了 WebWork 的设计核心,而不是 Struts 1 的设计核心。Struts 2 大量使用拦截器来处理用户的请求,从而允许用户的业务逻辑控制器和 Servlet API 分离。在处理请求的过程中以用户的业务逻辑控制器为目标,创建一个控制器代理,控制代理回调业务控制器中的 execute 方法来处理用户的请求,该方法的返回值决定了 Struts 2 以怎样的视图资源呈现给用户,如图 1.19 所示。

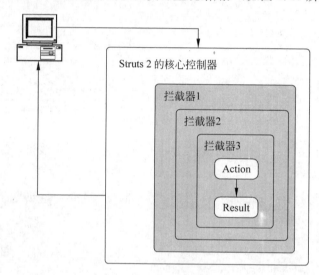

图 1.19 Struts 2 体系结构

从体系结构图 1.19 中可以发现 Struts 2 的大致处理流程如下。

(1) 浏览器发送请求,例如 admin/login.action。

(2) 核心控制器 StrutsPrepareAndExecuteFilter 根据请求调用对应的 Action。

(3) Struts 2 的拦截器链自动对请求进行相关应用的拦截,例如验证、数据类型转换或文件上传下载等功能。

(4) 回调 Action 的 execute 方法,该方法先获取参数,然后执行某种业务操作,既可以把数据保存到数据库中,也可以把数据从数据库中提取出来。实际上,Action 只是一个控制器,它会调用业务逻辑组件来处理用户的请求。

(5) Action 中 execute 方法的处理结果存入 Stack Context 中,并返回一个字符串。核心控制器 StrutsPrepareAndExecuteFilter 将根据返回的字符串跳转到指定的视图资源,该视图资源将会读取 Stack Context 中的信息,并在浏览器中显示信息。

### 1.2.2 能力目标

(1) 使用 Eclipse 或 MyEclipse 创建 Struts 2 应用。

(2) 发布 Struts 2 应用到 Tomcat 服务器并运行。

## 1.2.3 任务驱动

任务的主要内容如下。

（1）创建 Struts 2 应用。

（2）创建用户请求页面。

（3）Action 实现。

（4）Action 配置。

（5）发布 Struts 2 应用到 Tomcat 并运行。

**1. 创建 Struts 2 应用**

本节分别使用 IDE 开发工具 Eclipse 和 MyEclipse 创建 Web 项目 HelloWorld，Web 项目 HelloWorld 在 Eclipse 中的目录结构如图 1.20 所示。在 Eclipse 中如何创建项目、如何添加 Struts 2 类库到"HelloWorld"的 lib 目录下以及如何在 web.xml 文件中配置核心过滤器，请参考 1.1.3 节中的内容。Web 项目 HelloWorld 在 MyEclipse 中的目录结构如图 1.21 所示。本节涉及的 Action 以及配置文件将在后续章节中详细介绍。

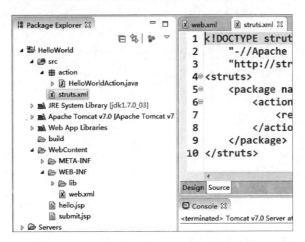

图 1.20　Web 项目 HelloWorld 在 Eclipse 中的目录结构

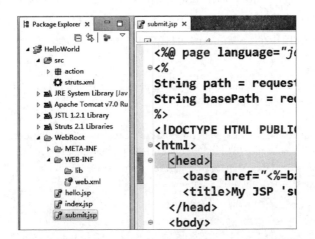

图 1.21　Web 项目 HelloWorld 在 MyEclipse 中的目录结构

### 2. 创建用户请求页面

在 HelloWorld 应用中，需要用户提交请求。submit.jsp 页面的代码如下：

```jsp
<%@ page language="java" contentType="text/html; charset=utf-8" pageEncoding="utf-8"%>
<!DOCTYPE html PUBLIC "-//W3C//DTD HTML 4.01 Transitional//EN" "http://www.w3.org/TR/html4/loose.dtd">
<html>
    <head>
        <meta http-equiv="Content-Type" content="text/html; charset=ISO-8859-1">
        <title>Insert title here</title>
    </head>
    <body>
        <form action="hello/world.action" method="post">
            <input type="submit" value="提交"/>
        </form>
    </body>
</html>
```

### 3. Action 实现

在"HelloWorld"应用的 src 目录下创建 HelloWorldAction 类来处理用户提交的请求，HelloWorldAction.java 的代码如下：

```java
package action;
public class HelloWorldAction {
    public String execute(){
        return "success";
    }
}
```

### 4. Action 配置

定义好 HelloWorldAction 后，还需要在 src 目录下提供一个 struts.xml 文件，并且需要在该 xml 文件中配置 HelloWorldAction。在 struts.xml 文件中主要配置处理请求的名称以及所对应的 Action 类，还需要配置 Action 处理结果和资源之间的映射关系。下面是 HelloWorldAction 应用中的 struts.xml 的代码：

```xml
<!DOCTYPE struts PUBLIC
    "-//Apache Software Foundation//DTD Struts Configuration 2.3//EN"
    "http://struts.apache.org/dtds/struts-2.3.dtd">
<struts>
    <package name="hello" namespace="/hello" extends="struts-default">
        <action name="world" class="action.HelloWorldAction">
            <result name="success">/hello.jsp</result>
        </action>
    </package>
</struts>
```

上面的映射文件定义了 name 为 world 的 Action，Action 将调用自身的 execute 方法处理用户请求,如果execute方法返回的结果为success字符串时,那么请求将转发到hello.jsp。hello.jsp的代码如下:

```
<%@ page language="java" contentType="text/html; charset=ISO-8859-1" pageEncoding="ISO-8859-1"%>
<!DOCTYPE html PUBLIC "-//W3C//DTD HTML 4.01 Transitional//EN" "http://www.w3.org/TR/html4/loose.dtd">
<html>
    <head>
        <meta http-equiv="Content-Type" content="text/html; charset=ISO-8859-1">
        <title>Insert title here</title>
    </head>
    <body>
        <h1>Hello World!</h1>
    </body>
</html>
```

**5. 发布Struts 2应用到Tomcat并运行**

1)在Eclipse中发布并运行Web应用

在Eclipse中第一次运行Web应用时,需要将Web应用发布到Tomcat。例如,运行HelloWorld应用时,可以选中jsp文件submit.jsp单击右键,选择Run As→Run on Server命令打开如图1.22所示的窗口,在窗口中单击Finish按钮即完成发布并运行。

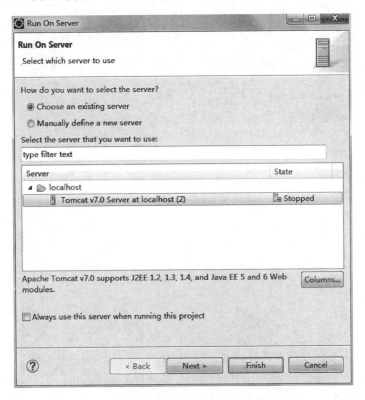

图1.22 在Eclipse中发布并运行Web应用

通过地址 http：//localhost：8080/HelloWorld/submit.jsp 首先访问 submit.jsp 页面，如图 1.23 所示。

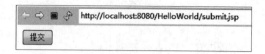

图 1.23　submit.jsp 页面

在如图 1.23 所示的页面中，用户单击"提交"按钮向 hello/world 发送请求，该请求被 StrutsPrepareAndExecuteFilter 拦截，并根据 struts.xml 文件中的配置转发给 HelloWorldAction 中的 execute 方法处理用户的请求，处理完成后，转发给 hello.jsp。hello.jsp 运行效果如图 1.24 所示。

2）在 MyEclipse 中发布并运行 Web 应用

在 MyEclipse 中发布并运行 Web 应用的步骤如下。

(1) 单击如图 1.25 所示方框中的图标，进入发布管理对话框，如图 1.26 所示。

图 1.24　hello.jsp 页面　　　　图 1.25　在 MyEclipse 发布 Web 应用

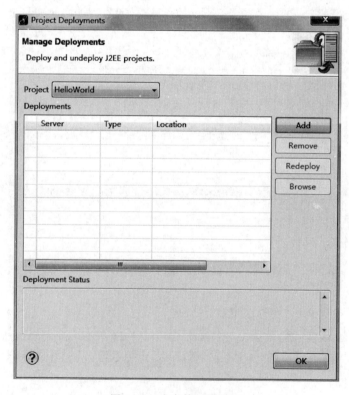

图 1.26　发布管理对话框

（2）在图 1.26 中单击 Add 按钮，打开服务器选择窗口，在该窗口的 Server 下拉列表中选择 Tomcat 7.x 选项如图 1.27 所示。

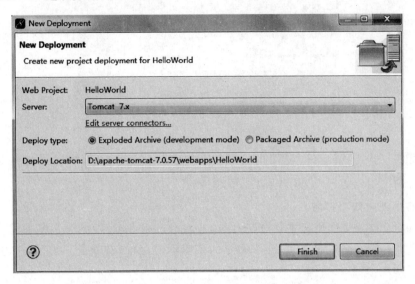

图 1.27　服务器选择窗口

（3）单击图 1.27 中的 Finish 按钮，发布完成。发布成功后，进入如图 1.28 所示的"发布管理"对话框。

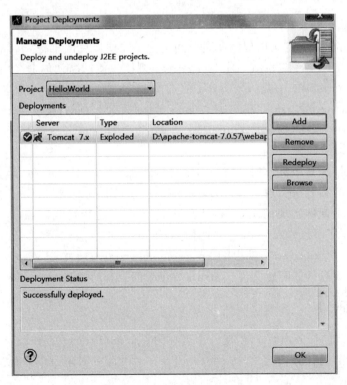

图 1.28　"发布管理"对话框

（4）启动 Tomcat。在 MyEclipse 中启动 Tomcat 如图 1.29 所示。

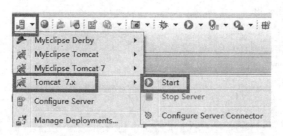

图 1.29　在 MyEclipse 中启动 Tomcat

（5）访问 Web 应用。在 MyEclipse 中，成功启动 Tomcat 后，可以通过地址 http://localhost:8080/HelloWorld/submit.jsp 访问 Web 应用。

#### 6. 任务小结或知识扩展

在 Struts 2 框架中使用包来管理 Action，包的作用和 Java 中的类包是非常类似的，它主要用于管理一组业务功能相关的 Action。在实际应用中，应该把一组业务功能相关的 Action 放在同一个包下。

1）name 属性

配置包时必须指定 name 属性，例如：本节 struts.xml 的包 hello。该 name 属性值可以任意取名，但必须唯一，它不对应 Java 的类包，如果其他包要继承该包，必须通过该属性进行引用。

2）namespace 属性

该属性是可选属性，定义该包的命名空间。为了方便以后的维护，namespace 最好用模块来进行命名。如果某个包没有指定 namespace，则该包使用默认的命名空间，默认的命名空间总是""。namespace 决定了 Action 的访问路径，默认为""，可以接收所有路径的 Action。假设 Action 的 name 为 world，namespace 写为 /、xxx 或者 /xxx/yyy，则对应的 Action 访问路径分别为 /world.action、/xxx/world.action 或者 /xxx/yyy/world.action。

3）extends 属性

该属性表示继承某个包。通常每个包都应该继承 struts-default 包，因为 Struts 2 很多核心的功能都是拦截器来实现。如：从请求中把请求参数封装到 Action、文件上传和数据验证等都是通过拦截器实现的。struts-default 包定义了这些拦截器和 Result 类型。可以这么说，当包继承了 struts-default 包才能使用 Struts 2 提供的核心功能。struts-default 包是在 struts2-core-2.x.x.jar 文件中的 struts-default.xml 文件中定义。struts-default.xml 文件也是 Struts 2 默认的配置文件，Struts 2 每次都会自动加载 struts-default.xml 文件。

### 1.2.4　实践环节

（1）参照本节的任务内容 1，创建一个名称为 sencondProject 的 Struts 2 应用。

（2）参照本节的任务内容 2，在 sencondProject 项目中创建一个名称为 yourFirst.jsp 的用户请求页面。

（3）参照本节的任务内容 3，实现 Action 类。

（4）参照本节的任务内容 4，配置 Action。

(5) 参照 1.2.3 小节的任务内容 5，发布 sencondProject 到 Tomcat 并运行。

## 1.3 本章小结

本章首先详细讲解了在 Eclipse 或 MyEclipse 中如何构建 Struts 2 开发环境；其次，以 HelloWorld 应用为例，简要介绍了 Struts 2 MVC 框架的基本流程。通过本章的学习，读者应该能够对 Struts 2 框架有个大致了解，并且能够独立开发 HelloWorld 应用实例。

开发 Struts 2 应用时，是使用 Eclipse，还是使用 MyEclipse，没有本质区别，应根据应用的实际需要选择合适的 IDE。本书以后章节的 Struts 2 应用都是使用 MyEclipse 开发的。

## 习 题 1

1. Struts 2 的核心过滤器是什么？如何配置该过滤器？
2. 简述 Struts 2 的体系结构。

# Struts 2 的 Action

**主要内容**

(1) Action 的作用。
(2) Action 的创建与配置。
(3) 动态方法调用以及通配符的使用。
(4) Action 接收用户数据的方式。
(5) Action 访问 Servlet API。
(6) 常见的 Result 类型。

在使用 Struts 2 进行 Web 应用开发时，Action 是应用的核心，Action 类包含了对用户请求的处理逻辑，因此被称为业务控制器。业务控制器 Action 是用户请求和业务逻辑之间的"桥梁"，是 Struts 2 框架最核心的部分，负责具体的业务逻辑处理。

## 2.1　Action 的创建与配置

对于 Struts 2 应用的开发者而言，Action 是应用的核心，开发者需要提供大量的 Action 类，并在 struts.xml 文件中配置 Action。

  2.1.1　核心知识

**1. Action 的创建方式**

在 Struts 2 应用中创建 Action 有 3 种方式：一个普通的类，不需要实现任何接口或继承任何类；实现 Action 接口；继承 ActionSupport 类，是最常用的一种方式，好处在于可以直接使用 Struts 2 封装好的方法。在这里以 1.2 节的 Action 类为例介绍 Action 的 3 种创建方式。

1) 一个普通的类

Struts 2 采用低侵入式设计，Struts 2 的 Action 可以不用继承或实现任何类或接口，代码重用性较好。1.2 节中的 HelloWorldAction 类就是一个普通的类，示例代码如下：

```
package action;
public class HelloWorldAction {
    public String execute(){
```

```
        return "success";
    }
}
```

2）实现 Action 接口

为了使开发者开发的 Action 更加规范，Struts 2 提供了一个 Action 接口，该接口定义了 Struts 2 的 Action 处理类应该实现的规范。标准 Action 接口的代码如下：

```
package com.opensymphony.xwork2;
public interface Action {
    //定义 Action 接口中包含的一些结果字符串
    public static final String SUCCESS = "success";
    public static final String NONE = "none";
    public static final String ERROR = "error";
    public static final String INPUT = "input";
    public static final String LOGIN = "login";
    //定义处理用户请求的 execute 方法
    public String execute() throws Exception;
}
```

在上面的 Action 接口代码中，只定义了一个 execute 方法，该方法返回一个字符串。除此之外，该接口还定义了 5 个字符串常量，它们的作用是统一 execute 方法的返回值。例如，当 Action 类处理用户请求成功后，有人喜欢返回 success 字符串，有人喜欢返回 doSuccess 字符串……这样不利于项目的统一管理。可以将 1.2 节中的 HelloWorldAction 修改为实现 Action 接口，代码如下所示：

```
package action;
import com.opensymphony.xwork2.Action;
public class HelloWorldAction implements Action{
    public String execute() throws Exception{
        return SUCCESS;
    }
}
```

在上面 HelloWorldAction 的代码中，execute 方法处理成功返回一个 SUCCESS 常量，不用直接返回一个 SUCCESS 字符串，这样统一了 execute 方法处理成功的返回值。

3）继承 ActionSupport 类

Struts 2 还提供了 Action 接口的一个实现类 ActionSupport。有关 ActionSupport 类的实现代码可以去 struts-2.x.x\src\xwork-core\src\main\java\com\opensymphony\xwork2 目录中查阅。ActionSupport 类是 Struts 2 默认的 Action 处理类，该类已经提供了许多默认方法，包括获取国际化信息方法、数据校验方法、默认的处理用户请求的方法等。如果让开发者的 Action 类继承该 ActionSupport 类，则会大大简化 Action 的开发。继承 ActionSupport 类创建 Action 的方式是最常用的一种方式。可以将 1.2 节中的 HelloWorldAction 修改为继承 ActionSupport 类，代码如下所示：

```
package action;
import com.opensymphony.xwork2.ActionSupport;
public class HelloWorldAction extends ActionSupport{
```

```
    public String execute(){
        return SUCCESS;
    }
}
```

### 2. Action 的配置

在 Struts 2 框架中每一个 Action 是一个工作单元。Struts 2 框架负责将一个用户请求对应到一个 Action 上去，每当一个 Action 类匹配一个请求的时候，这个 Action 类就会被 Struts 2 框架调用。struts.xml 中的每一对 action 元素就是一个 Action 映射。

Struts 2 框架提供多种 Action 配置方法，使得开发者有更多的选择。但在实际的项目开发过程中，开发团队必须采用统一的方法来开发，方便以后的维护。

例如，在 Struts 2 的 Web 应用 struts.xml 中，有如下配置：

```xml
<package name="user" namespace="/" extends="struts-default">
    <action name="login" class="action.LoginAction">
        <result>/loginSuccess.jsp</result>
        <result name="fail">/login.jsp</result>
    </action>
</package>
```

从上述的配置中，读者可以得知：类 LoginAction 提供一个 execute 方法，用来处理用户登录(login)的请求。Action 的名字是 login，用户请求此 Action 时需要将 URL 的访问路径设置为 http://localhost:8080/ch2/login.action。现在的问题是类 LoginAction 只能配置一个 Action 吗？如果将注册(register)也放在类 LoginAction 中，应该怎样配置呢？

Struts 2 配置文件中的 action 元素提供了一个名为 method 的属性，该属性指定执行 Action 时调用 Action 类中的任意公有、非静态的方法。那么登录和注册功能可以映射为：

```xml
<package name="user" namespace="/" extends="struts-default">
    <action name="login" class="action.LoginAction">
        <result>/loginSuccess.jsp</result><!-- 与 execute 方法的 return SUCCESS 对应 -->
        <result name="loginfail">/login.jsp</result><!-- 与 execute 方法的 return "loginfail"对应 -->
    </action>
    <action name="register" class="action.LoginAction" method="register">
        <result>/login.jsp</result><!-- 与 register 方法的 return SUCCESS 对应 -->
        <result name="registerfail">/register.jsp</result><!-- return "registerfail"对应 -->
    </action>
</package>
```

从上述的配置中，读者可以得知：类 LoginAction 提供一个 register 方法，用来处理用户注册的请求。Action 的名字是 register，用户请求此 Action 时需要将 URL 的访问路径设置为 http://localhost:8080/ch2/register.action。

类 LoginAction 的伪代码如下：

```
public class LoginAction extends ActionSupport{
/**
```

```java
 * 实现登录功能
 */
public String execute(){
    if(登录成功){
        return SUCCESS;
    }
    return "loginfail";        //与配置文件中result的name属性对应
}
/**
 * 实现注册功能
 */
public String register(){
    if(注册成功){
        return SUCCESS;
    }
    return "registerfail";     //与配置文件中result的name属性对应
}
}
```

#### 3. DMI 以及通配符

假设有这样一个 Action 类 UserAction 实现对用户 User 的 crud(增删改查)功能 add()、del()、modify()、query()，按照前面的学习，在 struts.xml 中配置 4 个 Action：

```xml
<package name="test1" namespace="/" extends="struts-default">
    <action name="addUser" class="action.UserAction" method="add">
        <result>/User/addUser.jsp</result>
    </action>
    <action name="delUser" class="action.UserAction" method="del">
        <result>/User/delUser.jsp</result>
    </action>
    <action name="modifyUser" class="action.UserAction" method="modify">
        <result>/User/modifyUser.jsp</result>
    </action>
    <action name="queryUser" class="action.UserAction" method="query">
        <result>/User/queryUser.jsp</result>
    </action>
</package>
```

然后在 index.jsp 页面中请求这 4 个 Action：

```html
<body>
    User:
    <a href="addUser.action">add</a>
    <a href="delUser.action">del</a>
    <a href="modifyUser.action">modify</a>
    <a href="queryUser.action">query</a>
</body>
```

设想一下如果有大量的 Action，配置文件就会变得很臃肿，为以后的维护埋下隐患。解决配置文件过于庞大的方法有：动态方法调用(Dynamic Method Invocation，DMI)和通配符配置。

1）DMI

DMI 的用法非常简单：在 Action 名后加上"!xxx"（xxx 为方法名）。在 index.jsp 页面可以这样请求这 4 个 Action：

```
<body>
    DMI User:
    <a href = "UserCRUD!add.action">add</a>
    <a href = "UserCRUD!del.action">del</a>
    <a href = "UserCRUD!modify.action">modify</a>
    <a href = "UserCRUD!query.action">query</a>
</body>
```

配置时，只需要配置一个 Action 即可：

```
<action name = "UserCRUD" class = "action.UserAction">
    <result>/User/User.jsp</result>
</action>
```

使用 DMI 有一个潜在的问题，对用户的增删改查，所使用的配置和验证信息都是上面这个配置，不够灵活。接下来的通配符就与 DMI 有着很大的区别。

2）通配符

通配符配置其实也只需要配置一个 Action 即可，但由于使用通配符，可动态地将 URL 请求与配置文件中的通配符结合，变成真正的 Action 配置。

注意对用户的增删改查，所有文件都采用统一的命名规范操作+User，所以可以找一个通用的符号代替这些操作——通配符（*）。对于这个例子可以将 struts.xml 文件配置如下：

```
<package name = "test1" namespace = "/" extends = "struts-default">
    <action name = "*User" class = "action.UserAction" method = "{1}">
        <result>/User/{1}User.jsp</result>
    </action>
</package>
```

上述配置中，* 表示一个通配符，{1}表示第一个通配符代表的含义。

### 2.1.2 能力目标

掌握 Action 的创建与配置，了解通配符的使用方法。

### 2.1.3 任务驱动

1）任务的主要内容

在 Struts 2 的 Web 应用 ch2 中，首先，通过继承 ActionSupport 类的方式创建一个 Action 类 UserAction，该 Action 类有两个逻辑处理方法：login 与 logout；其次，编写一个 JSP 页面 task_2_1.jsp，在该 JSP 页面中有两个表单，每个表单分别有一个提交按钮，两个按钮提交请求分别由 Action 类的方法 login 与 logout 进行处理，处理结果只有成功；最后，在 struts.xml 中配置 Action，login 成功跳转到 loginSuccess.jsp，logout 成功跳转到

login.jsp。

2) 任务的代码模板

UserAction.java 的代码模板如下：

```
package action;
import com.opensymphony.xwork2.ActionSupport;
public class UserAction extends【代码1】{
    private static final long serialVersionUID = 1L;
    /**
     * 处理登录请求
     */
    public String【代码2】{
        return SUCCESS;
    }
    /**
     * 处理安全退出请求
     */
    public String【代码3】{
        return SUCCESS;
    }
}
```

task_2_1.jsp 的代码如下：

```
<%@ page language="java" import="java.util.*" pageEncoding="UTF-8"%>
<%
  String path = request.getContextPath();
  String basePath = request.getScheme() + "://" + request.getServerName() + ":" + request.getServerPort() + path + "/";
%>
<!DOCTYPE HTML PUBLIC "-//W3C//DTD HTML 4.01 Transitional//EN">
<html>
  <head>
    <base href="<%=basePath%>">
    <title>My JSP 'task_2_1.jsp' starting page</title>
  </head>
  <body>
    <form action="user/login.action">
    <input type="submit" value="登录">
    </form>
    <form action="user/logout.action">
    <input type="submit" value="安全退出">
    </form>
  </body>
</html>
```

struts.xml 的代码模板如下：

```
…
<package name="myuser" namespace=【代码4】extends="struts-default">
    <action name="login" class="action.UserAction" method=【代码5】>
```

```
                <result>/loginSuccess.jsp</result>
            </action>
            <action name=【代码 6】 class="action.UserAction" method="logout">
                <result>/login.jsp</result>
            </action>
        </package>
     ……
```

loginSuccess.jsp 与 login.jsp 的代码略。

3) 任务小结或知识扩展

在 xml 配置文件中,如果没有为 Action 指定 class 属性,默认是 ActionSupport 类。如果没有为 Action 指定 method 属性,默认执行 Action 中的 execute 方法。如果没有指定 result 的 name 属性,默认值为 success。

4) 任务代码模板的参考答案

【代码 1】`ActionSupport`

【代码 2】`login()`

【代码 3】`logout()`

【代码 4】`"/user"`

【代码 5】`"login"`

【代码 6】`"logout"`

### 2.1.4 实践环节

在 Struts 2 的 Web 应用 ch2 中,创建一个 Action 类 ComputerAction,该类有 add()、del()、modify()、query() 4 个方法对 Computer 进行增删改查。在 add()、del()、modify()、query() 4 个方法中只有"return SUCCESS;"语句,再编写一个 JSP 文件 computer.jsp,在该 JSP 文件中请求 4 个 Action,JSP 文件代码如下:

```
<body>
    Computer:
        <a href="addComputer.action">add</a>
        <a href="delComputer.action">del</a>
        <a href="modifyComputer.action">modify</a>
        <a href="queryComputer.action">query</a>
</body>
```

请在 struts.xml 中使用通配符配置这 4 个 Action,并测试运行。

## 2.2　Action 接收请求参数

如何将用户的请求信息提交给 Action 是 Struts 2 框架的重点知识,因为 Action 只有获取了用户的请求信息后,才能对请求做出处理。在 Struts 2 的 Action 中,有 3 种方式接收用户的请求参数,分别是使用 Action 属性接收参数、使用 DomainModel(实体)接收参数和使用 ModelDriven<T>接口接收参数。本节使用同一个完整示例讲解 Action 接收请求参数的 3 种方式。

## 2.2.1 核心知识

**1. 使用 Action 属性接收参数**

使用 Action 属性接收参数就是在 Action 类中定义与请求参数同名的属性,并为属性提供 setter 和 getter 方法,Struts 2 便能自动接收请求参数并赋值给同名属性。下面通过一个完整的示例来讲解 Action 接收请求参数。首先,提供一个 input.jsp 页面,代码如下:

```html
<body>
    <form action="input.action" method="post">
        <h3>信息提交</h3>
        用户名:<input type="text" name="uname"><br>
        年龄:<input type="text" name="uage"><br>
        <input type="submit" value="提交">
    </form>
</body>
```

其次,需要提供一个 Action 类,该 Action 是通过属性来保存用户请求的参数,要保证 input.jsp 页面 form 表单中元素的 name 属性与 Action 类中的属性名称一致,ReceiveAction 代码如下:

```java
package action;
import com.opensymphony.xwork2.ActionSupport;
public class ReceiveAction extends ActionSupport{
    //用来封装用户请求中用户名的信息
    private String uname;
    //用来封装用户请求中年龄的信息
    private int uage;
    public String getUname() {
        return uname;
    }
    public void setUname(String uname) {
        this.uname = uname;
    }
    public int getUage() {
        return uage;
    }
    public void setUage(int uage) {
        this.uage = uage;
    }
    /**
     * 处理用户请求的方法
     */
    public String execute(){
        System.out.println("uname:" + uname);
        System.out.println("uage:" + uage);
        return SUCCESS;
    }
}
```

## 2. 使用 DomainModel(实体)接收参数

使用 DomainModel 接收参数就是将用户请求的参数封装到 JavaBean 对象中，所以首先提供一个 JavaBean 类 Information，代码如下：

```java
package model;
public class Information {
    //用来封装用户请求中用户名的信息，请求参数 uname 与属性名相同
    private String uname;
    //用来封装用户请求中年龄的信息，请求参数 uage 与属性名相同
    private int uage;
    public String getUname() {
        return uname;
    }
    public void setUname(String uname) {
        this.uname = uname;
    }
    public int getUage() {
        return uage;
    }
    public void setUage(int uage) {
        this.uage = uage;
    }
}
```

其次，提供处理请求的 ReceiveAction，代码如下：

```java
package action;
import model.Information;
import com.opensymphony.xwork2.ActionSupport;
public class ReceiveAction extends ActionSupport{
    //声明实体(JavaBean)对象 info，提供 setter 和 getter 方法
    private Information info;
    public Information getInfo() {
        return info;
    }
    public void setInfo(Information info) {
        this.info = info;
    }
    /**
     * 处理用户请求的方法
     */
    public String execute(){
        System.out.println("uname:" + info.getUname());
        System.out.println("uage:" + info.getUage());
        return SUCCESS;
    }
}
```

最后，input.jsp 页面 form 表单中元素的 name 属性值需要加上实体对象名"info"，代码如下：

```html
<body>
    <form action = "input.action" method = "post">
        <h3>信息提交</h3>
        用户名:<input type = "text" name = "info.uname"><br>
        年龄:<input type = "text" name = "info.uage"><br>
        <input type = "submit" value = "提交">
    </form>
</body>
```

### 3. 使用 ModelDriven<T>接口接收参数

使用 ModelDriven<T>接口接收参数时,Action 类需要实现 ModelDriven<T>接口,T 为 JavaBean 类。ReceiveAction 的代码如下:

```java
package action;
import model.Information;
import com.opensymphony.xwork2.ActionSupport;
import com.opensymphony.xwork2.ModelDriven;
public class ReceiveAction extends ActionSupport implements ModelDriven<Information>{
    //定义 info 对象,需要创建该对象,不会主动帮你创建
    private Information info = new Information();
    /**
     * 处理用户请求的方法
     */
    public String execute(){
        System.out.println("uname:" + info.getUname());
        System.out.println("uage:" + info.getUage());
        return SUCCESS;
    }
    @Override
    public Information getModel() {
        return info;
    }
}
```

input.jsp 页面 form 表单中元素的 name 属性值与 JavaBean 类中的属性名称一致,代码如下:

```html
<body>
    <form action = "input.action" method = "post">
        <h3>信息提交</h3>
        用户名:<input type = "text" name = "uname"><br>
        年龄:<input type = "text" name = "uage"><br>
        <input type = "submit" value = "提交">
    </form>
</body>
```

## 2.2.2 能力目标

掌握 Action 接收请求参数的 3 种方式。

## 2.2.3 任务驱动

**1）任务的主要内容**

编写一个登录页面 myLogin.jsp，在该页面中有用户名和密码两个输入框，输入登录信息后，将信息提交给 MyLoginAction 类处理，该 Action 使用 DomainModel（实体类为 MyUser）方式接收用户名和密码。当用户名为"陈恒"、密码为"123456"时登录成功，否则失败。登录成功时显示 myLoginSuccess.jsp 页面，登录失败时显示 myLoginFail.jsp 页面。

**2）任务的代码模板**

myLogin.jsp 的代码如下：

```jsp
<%@ page language = "java" import = "java.util.*" pageEncoding = "UTF-8" %>
<%
    String path = request.getContextPath();
    String basePath = request.getScheme() + "://" + request.getServerName() + ":" + request.getServerPort() + path + "/";
%>
<!DOCTYPE HTML PUBLIC " - //W3C//DTD HTML 4.01 Transitional//EN">
<html>
  <head>
    <base href = "<% = basePath %>">
    <title>My JSP 'myLogin.jsp' starting page</title>
  </head>
  <body>
    <form action = "myLogin.action" method = "post">
        用户名：<input type = "text" name = "mu.userName"><br>
        密码：<input type = "password" name = "mu.userpwd"><br>
        <input type = "submit" value = "提交">
    </form>
  </body>
</html>
```

MyUser.java 的代码如下：

```java
package model;
public class MyUser {
    private String userName;
    private String userpwd;
    public String getUserName() {
        return userName;
    }
    public void setUserName(String userName) {
        this.userName = userName;
    }
    public String getUserpwd() {
        return userpwd;
    }
    public void setUserpwd(String userpwd) {
        this.userpwd = userpwd;
    }
}
```

MyLoginAction.java 的代码模板如下：

```java
package action;
import model.MyUser;
import com.opensymphony.xwork2.ActionSupport;
public class MyLoginAction extends ActionSupport{
    private static final long serialVersionUID = 1L;
    //声明 MyUser 的对象 mu
    【代码 1】
    public String execute(){
        //登录成功
        if("陈恒".equals(mu.getUserName()) && "123456".equals(mu.getUserpwd())){
            return SUCCESS;
        }
        return "loginFail";
    }
    public MyUser getMu() {
        return mu;
    }
    public void setMu(MyUser mu) {
        this.mu = mu;
    }
}
```

struts.xml 的代码模板如下：

```xml
…
<package name = "youruser" namespace = "/" extends = "struts-default">
    <action【代码 2】class = "action.MyLoginAction">
        <result>/myLoginSuccess.jsp</result>
        <result【代码 3】>/myLoginFail.jsp</result>
    </action>
</package>
…
```

myLoginSuccess.jsp 与 myLoginFail.jsp 的代码略。

3）任务小结或知识扩展

通过学习 Action 接收请求参数的 3 种方式，可以发现各有优点，开发者可根据实际情况灵活使用。

4）任务代码模板的参考答案

【代码 1】 private MyUser mu;

【代码 2】 name = "myLogin"

【代码 3】 name = "loginFail"

## 2.2.4 实践环节

将 2.2.3 节中"MyLoginAction 类接收请求参数的方式"修改为"使用 ModelDriven<T>接口方式接收请求参数"。

## 2.3 Action 访问 Servlet API

在 Struts 2 中，Action 已经与 Servlet API 分离，使得 Action 具有更加灵活和低耦合的特性。Action 作为 Web 应用的控制器，不访问 Servlet API 几乎是不可能的，例如跟踪 Http-Session 状态等。Web 应用中通常需要访问的 Servlet API 是 HttpServletRequest、HttpSession 和 ServletContext，这 3 个类分别代表 request、session 和 application 等 JSP 内置对象。

### 2.3.1 核心知识

Action 访问 Servlet API 的常用方式有通过 ActionContext，实现 RequestAware、SessionAware 和 ApplicationAware 接口，通过 ServletActionContext 以及实现 ServletRequestAware 接口等方式。

#### 1. 通过 ActionContext

Struts 2 提供了一个 ActionContext 类，在 Action 中可以通过该类访问 Servlet API。ActionContext 类中包含的几个常用方法如表 2.1 所示。

表 2.1 ActionContext 类的常用方法

| 序号 | 方　　法 | 功　能　说　明 |
| --- | --- | --- |
| 1 | Object get(Object key) | 通过参数 key 获取当前 ActionContext 中的值 |
| 2 | Map getApplication( ) | 返回一个 Map 类型的 Application 对象 |
| 3 | static ActionContext getContext( ) | 返回当前线程的 ActionContext 对象 |
| 4 | Map getParameters( ) | 返回一个包含所有 HttpServletRequest 参数信息的 Map 对象 |
| 5 | Map getSession( ) | 返回一个 Map 类型的 HttpSession 对象 |

下面的 Web 应用将在 Action 中通过 ActionContext 访问 Servlet API，以图 2.1 所示的登录页面为示例。yourLogin.jsp 的代码如下：

```
<body>
    <h3>登录页面</h3>
    <form action = "yourLogin.action" method = "post">
        <table>
            <tr>
                <td>用户名：</td>
                <td><input type = "text" name = "userName"/></td>
            </tr>
            <tr>
                <td>密　码：</td>
                <td><input type = "password" name = "userpwd"/></td>
            </tr>
            <tr>
                <td><input type = "submit" value = "提交"/></td>
                <td><input type = "reset" value = "重置"/></td>
```

```
        </tr>
    </table>
</form>
</body>
```

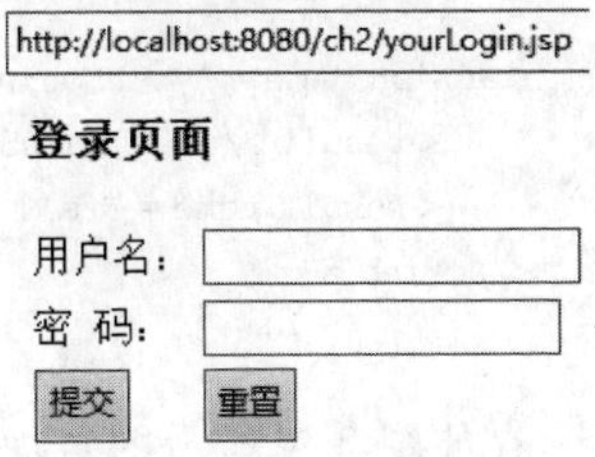

当单击如图2.1所示页面中的"提交"按钮时,系统将提交到yourLogin.action,该Action对应的处理类YourLoginAction的代码如下:

图2.1 简单的登录页面

```java
package action;
import java.util.Map;
import model.MyUser;
import com.opensymphony.xwork2.ActionContext;
import com.opensymphony.xwork2.ActionSupport;
import com.opensymphony.xwork2.ModelDriven;
public class YourLoginAction extends ActionSupport implements ModelDriven<MyUser>{
    private static final long serialVersionUID = 1L;
    // MyUser 为 JavaBean 类
    private MyUser u = new MyUser();
    //声明 map 对象
    private Map<String,Object> request;
    private Map<String,Object> session;
    private Map<String,Object> application;
    //构造方法,初始化 map 对象
    public YourLoginAction(){
        //获取 ActionContext 实例,通过该实例访问 Servlet API
        ActionContext ctx = ActionContext.getContext();
        request = (Map<String,Object>)ctx.get("request");
        session = ctx.getSession();
        application = ctx.getApplication();
    }
    public String execute(){
        if("陈恒".equals(u.getUserName()) && "123456".equals(u.getUserpwd())){
            session.put("usersession", u);
            application.put("userapplication", u);
            return SUCCESS;     //跳转到 yourLoginSuccess.jsp 页面
        }else{
            //类似 request.setAtrribute("userrequest",u);
            request.put("userrequest", u);
            return "fail";      //跳转到 yourLoginFail.jsp 页面
        }
    }
    @Override
    public MyUser getModel() {
        // TODO Auto-generated method stub
        return u;
    }
}
```

上面的Action访问了HttpServletRequest、HttpSession和ServletContext。将该Action配置在struts.xml文件中,具体代码如下:

```
...
<action name = "yourLogin" class = "action.YourLoginAction">
    <result>/yourLoginSuccess.jsp</result>
    <result name = "fail">/yourLoginFail.jsp</result>
</action>
...
```

在如图 2.1 所示的页面中输入用户名为"陈恒",密码为"123456"时,单击"提交"按钮跳转到 yourLoginSuccess.jsp 页面,否则跳转到 yourLoginFail.jsp 页面。

yourLoginSuccess.jsp 页面的代码如下:

```
<body>
    从 session 中取值<br>
    用户名:${sessionScope.usersession.userName}<br>
    密码:${sessionScope.usersession.userpwd}<br>
    从 appication 中取值<br>
    用户名:${applicationScope.userapplication.userName}<br>
    密码:${applicationScope.userapplication.userpwd}
</body>
```

yourLoginFail.jsp 页面的代码如下:

```
<body>
    从 request 中取值<br>
    用户名:${requestScope.userrequest.userName}<br>
    密码:${requestScope.userrequest.userpwd}
</body>
```

### 2. 实现 RequestAware、SessionAware 和 ApplicationAware 接口

在 Struts 2 的 Action 中还可以通过实现一系列接口 RequestAware、SessionAware 和 ApplicationAware 来访问 Servlet API。现在只需要将 Action 类 YourLoginAction 修改为如下代码:

```
package action;
import java.util.Map;
import model.MyUser;
import org.apache.struts2.interceptor.ApplicationAware;
import org.apache.struts2.interceptor.RequestAware;
import org.apache.struts2.interceptor.SessionAware;
import com.opensymphony.xwork2.ActionSupport;
import com.opensymphony.xwork2.ModelDriven;
public class YourLoginAction extends ActionSupport implements ModelDriven<MyUser>,
RequestAware,SessionAware,ApplicationAware{
    private static final long serialVersionUID = 1L;
    private MyUser u = new MyUser();
    //声明 map 对象
    private Map<String,Object> request;
    private Map<String,Object> session;
    private Map<String,Object> application;
    public String execute(){
```

```java
            if("陈恒".equals(u.getUserName()) && "123456".equals(u.getUserpwd())){
                session.put("usersession", u);
                application.put("userapplication", u);
                return SUCCESS;
            }else{
                //类似 request.setAtrribute("userrequest",u);
                request.put("userrequest", u);
                return "fail";
            }
        }
        @Override
        public MyUser getModel() {
            return u;
        }
        @Override
        public void setSession(Map<String, Object> arg0) {
            session = arg0;
        }
        @Override
        public void setRequest(Map<String, Object> arg0) {
            request = arg0;
        }
        @Override
        public void setApplication(Map<String, Object> arg0) {
            application = arg0;
        }
    }
```

### 3. 通过 ServletActionContext

在 Struts 2 的 Action 中，除了前面两种间接方式访问 Servlet API 之外，还可以直接访问 Servlet API。直接访问 Servlet API 有两种方式：通过 ServletActionContext 和实现 ServletRequestAware 接口。

通过 ServletActionContext 直接访问 Servlet API，只需要将 Action 类 YourLoginAction 修改为如下代码：

```java
package action;
import javax.servlet.ServletContext;
import javax.servlet.http.HttpServletRequest;
import javax.servlet.http.HttpSession;
import org.apache.struts2.ServletActionContext;
import model.MyUser;
import com.opensymphony.xwork2.ActionSupport;
import com.opensymphony.xwork2.ModelDriven;
public class YourLoginAction extends ActionSupport implements ModelDriven<MyUser> {
    private static final long serialVersionUID = 1L;
    private MyUser u = new MyUser();
    private HttpServletRequest request;
    private HttpSession session;
    private ServletContext application;
```

```java
public YourLoginAction() {
    request = ServletActionContext.getRequest();
    session = request.getSession();
    application = ServletActionContext.getServletContext();
}
public String execute() {
    if ("陈恒".equals(u.getUserName()) && "123456".equals(u.getUserpwd())) {
        session.setAttribute("usersession", u);
        application.setAttribute("userapplication", u);
        return SUCCESS;
    } else {
        request.setAttribute("userrequest", u);
        return "fail";
    }
}
@Override
public MyUser getModel() {
    return u;
}
}
```

### 4. 实现 ServletRequestAware 接口

实现 ServletRequestAware 接口直接访问 Servlet API，只需要将 Action 类 YourLoginAction 修改为如下代码：

```java
package action;
import javax.servlet.ServletContext;
import javax.servlet.http.HttpServletRequest;
import javax.servlet.http.HttpSession;
import model.MyUser;
import org.apache.struts2.interceptor.ServletRequestAware;
import com.opensymphony.xwork2.ActionSupport;
import com.opensymphony.xwork2.ModelDriven;
public class YourLoginAction extends ActionSupport implements ModelDriven<MyUser>,ServletRequestAware{
    private static final long serialVersionUID = 1L;
    private MyUser u = new MyUser();
    private HttpServletRequest request;
    private HttpSession session;
    private ServletContext application;
    public String execute(){
        if("陈恒".equals(u.getUserName()) && "123456".equals(u.getUserpwd())){
            session.setAttribute("usersession", u);
            application.setAttribute("userapplication", u);
            return SUCCESS;
        }else{
            request.setAttribute("userrequest", u);
            return "fail";
        }
    }
}
```

```
        @Override
        public MyUser getModel() {
            return u;
        }
        @Override
        public void setServletRequest(HttpServletRequest arg0) {
            request = arg0;
            session = arg0.getSession();
            application = arg0.getServletContext();
        }
    }
```

### 2.3.2 能力目标

掌握 Action 访问 Servlet API 的常用方式。

### 2.3.3 任务驱动

1) 任务的主要内容

编写一个 Action 类 CountNumberAction，在该 Action 类中使用"实现 ApplicationAware 接口方式"统计访问该 Action 的次数。首先，通过 http://localhost:8080/ch2/countNumber.action 访问该 Action；然后，访问成功后跳转到 showNumber.jsp 页面显示统计结果。

2) 任务的代码模板

CountNumberAction.java 的代码模板如下：

```
package action;
import java.util.Map;
import org.apache.struts2.interceptor.ApplicationAware;
import com.opensymphony.xwork2.ActionSupport;
public class CountNumberAction extends ActionSupport implements 【代码1】{
    private static final long serialVersionUID = 1L;
    Map<String, Object> application;
    public String execute(){
        int number = 1;
        Integer count = (Integer)application.get("countNumber");
        if(count != null){        //不是第一个访客
            count++;
            number = count;
        }
        【代码2】                    //将 number 以 countNumber 为关键字保存在 application 对象中
        return SUCCESS;
    }
    @Override
    public void setApplication(Map<String, Object> arg0) {
        【代码3】
    }
}
```

showNumber.jsp 的代码如下：

```jsp
<%@ page language="java" import="java.util.*" pageEncoding="UTF-8"%>
<%
    String path = request.getContextPath();
    String basePath =
    request.getScheme()+"://"+request.getServerName()+":"+request.getServerPort()+path+"/";
%>
<!DOCTYPE HTML PUBLIC "-//W3C//DTD HTML 4.01 Transitional//EN">
<html>
  <head>
    <base href="<%=basePath%>">
    <title>My JSP 'showNumber.jsp' starting page</title>
  </head>
  <body>
    您是第${applicationScope.countNumber}个访问 CountNumberAction 的人。
  </body>
</html>
```

Action 的配置文件内容略。

3）任务小结或知识扩展

在 Action 中访问 Servlet API，本节介绍了 4 种常用方式。其中，"通过 ActionContext"与"实现 RequestAware、SessionAware 和 ApplicationAware 接口"这两种方式与 Servlet 进行了解耦。在实际编程中，具体使用哪种方式，根据程序需要进行选择。本书推荐使用"实现 RequestAware、SessionAware 和 ApplicationAware 接口"的方式访问 Servlet API。

4）任务代码模板的参考答案

【代码1】ApplicationAware

【代码2】application.put("countNumber", number);

【代码3】application = arg0;

## 2.3.4 实践环节

创建 Struts 2 Web 应用 project233practice，具体要求如下。

（1）编写登录页面 login.jsp，显示登录表单（账号，密码，提交按钮），提交请求到登录处理 Action。

（2）编写登录处理 action.LoginAction，在该 Action 中：

① 取得提交的账号和密码。

② 如果账号或密码任何一个为空，返回到登录页面。

③ 如果账号和密码均不为空，通过 ActionContext 方式将账号保存到会话对象 session 中。

④ 使用 ServletContext 对象 application，统计在线用户的个数。

⑤ 跳转到系统主页 main.jsp。

（3）编写系统主页 main.jsp，显示保存在 session 中的用户账号，显示在线用户人数，注销链接，单击注销请求到注销处理 Action。

（4）编写注销处理 action.LogoutAction，在该 Action 中。

① 使用实现 SessionAware 接口方式清除 session 对象。

② 使用 ServletContext 对象 application 完成在线用户人数的减少功能。
③ 跳转到登录页面 login.jsp。

## 2.4 Action 中常见的结果类型

Struts 2 的 Action 处理用户请求结束后,返回一个普通字符串——逻辑视图名,必须在配置文件中完成逻辑视图和物理视图资源的映射,才可让系统转到实际的视图资源。视图资源告诉用户系统执行的情况。但有的时候,返回的仅仅是个 xml 文件、json 串或输出流之类的,于是原有的代码很可能会因不同的返回结果而进行修改。Struts 2 提供了可配置的方式来帮助用户完成上述需求,用户不需要手动处理各种返回结果的细节。需要做的仅仅是在配置文件配置所希望的返回结果类型。

### 2.4.1 核心知识

**1. chain 类型**

chain 结果类型是 Action 链式处理的结果类型,如下两个 Action 的配置:

```
...
<action name="addUser" class="action.UserAction" method="add">
    <result type="chain">queryUser</result>
    <result name="fail">/User/addUser.jsp</result>
</action>
<action name="queryUser" class="action.UserAction" method="query">
    <result>/User/queryUser.jsp</result>
</action>
...
```

上面的配置可以这样理解,成功执行名为 addUser 的 Action 后,直接使用 chain 类型跳转到同一个包中名为 queryUser 的 Action 中。这里的包是配置文件中的包。

如果需要从一个 Action 中跳转到不同包的 Action 中,该如何配置呢?示例代码如下所示:

```
<package name="test1" namespace="/test1" extends="struts-default">
    <action name="addUser" class="action.UserAction" method="add">
        <result type="chain">
            <!-- actionName 为固定写法,queryUser 为 Action 的名字 -->
            <param name="actionName">queryUser</param>
            <!-- namespace 为固定写法,test2 为包的 namespace -->
            <param name="namespace">/test2</param>
        </result>
        <result name="fail">/User/addUser.jsp</result>
    </action>
    ...
</package>
...
<package name="test2" namespace="/test2" extends="struts-default">
    <action name="queryUser" class="action.UserAction" method="query">
```

```xml
            <result>/User/queryUser.jsp</result>
        </action>
        …
</package>
```

需要注意的是，chain 类型会将第一个 Action 执行完的状态复制到第二个 Action 中，供其使用。这样，在第二个 Action 中可以直接使用或修改第一个 Action 中的值，所以在最终页面显示的时候，很可能结果是不正确的。另外，chain 类型，也不能实现跳转时通过"?"传递参数。因此，不推荐使用 chain 结果类型。

如果需要从一个 Action 中跳转到不同包的 Action 中，并通过"?"传递参数，又该如何配置呢？这时可以使用 redirectAction 类型或 redirect 类型。

### 2. redirectAction 类型

当需要让一个 Action 处理结束后，直接将请求重定向（请求参数、属性和处理结果都将丢失）到另一个 Action 时，应该使用 redirectAction 类型。

配置 redirectAction 结果类型时，可以指定如下两个参数。

（1）actionName：该参数指定重定向的 Action 名。

（2）namespace：该参数指定重定向的 Action 所在包的命名空间。

例如，需要从一个 Action 中跳转到不同包的 Action 中，并通过"?"传递参数，示例代码如下：

```xml
<package name="test1" namespace="/test1" extends="struts-default">
    <action name="addUser" class="action.UserAction" method="add">
        <result type="redirectAction">
            <!-- 由于是 xml 文档,传参时必须将"&"写成"&amp;",id 和 uname 是参数名,idv 和 unamev 是 Action 中的属性名,${ } 为 EL 表达式. -->
            <param name="actionName">queryUser?id=${idv}&amp;uname=${unamev}</param>
            <param name="namespace">/test2</param>
        </result>
        <result name="fail">/User/addUser.jsp</result>
    </action>
    …
</package>
…
<package name="test2" namespace="/test2" extends="struts-default">
    <action name="queryUser" class="action.UserAction" method="query">
        <result>/User/queryUser.jsp</result>
    </action>
    …
</package>
```

### 3. redirect 类型

redirect 类型与 redirectAction 类型不同的是，既可以重定向到 JSP 页面，也可以重定向到 Action，甚至可以重定向到其他网站，这种配置比较常用。

配置 redirect 结果类型时，可以指定参数 location，该参数指定 Action 处理完用户请求后跳转的 Action 地址。

示例代码如下：

```xml
<package name="test1" namespace="/test1" extends="struts-default">
    <action name="addUser" class="action.UserAction" method="add">
        <result type="redirect">
            <!-- 同一个包时，location 的参数值直接写 xxx.action；不同包时，加上另一个包的 namespace，如：/xxx/yyy.action -->
            <param name="location">queryUser.action</param>
            <!-- 直接设置参数，'query'为字符串常量，请求 addUser 成功后，地址栏显示：
            http://localhost:8080/ch2/queryUser.action?act=query -->
            <param name="act">${'query'}</param>
        </result>
        <!-- 上面的 result 还可以这样简单配置（推荐）
        <result type="redirect">queryUser.action?act=${'query'}</result> (在 MyEclipse 2014 中，此处代码会出现错误提示，但不影响程序正常运行。关闭 MyEclipse 2014，再打开时，此错误提示消除。)
        -->
        <!-- 也可以通过?设置参数，请求 addUser 失败后，地址栏显示：
        http://localhost:8080/ch2/User/addUser.jsp?act=add -->
        <result type="redirect" name="addfail">/User/addUser.jsp?act=${'add'}</result>
    </action>
    …
    <action name="queryUser" class="action.UserAction" method="query">
        <result>/User/queryUser.jsp</result>
    </action>
    …
</package>
```

### 4. dispatcher 类型

dispatcher 结果类型是将请求转发到指定的 JSP 资源，在目标页面中可以访问 request 作用域的数据。默认结果类型就是 dispatcher 类型，即 result 没有指定 type 属性时，该 result 类型就是 dispatcher 类型。

### 5. 全局 result

一般情况下，<result>节点都配置在<action>中，作为 Action 的返回结果。但有些情况下，这种做法却会造成冗余的配置。比如说，系统只有在用户合法登录的情况下才能继续操作，否则应该跳转到登录页面。这时候，就可以使用全局 result 了。

Struts 2 允许在<package>节点中使用<global-results>定义全局结果。当 Action 执行完成后，首先会从自己的<result>节点中匹配，如果匹配不成功，再从定义好的全局结果中继续查找。示例代码如下：

```xml
<package name="test1" namespace="/test1" extends="struts-default">
…
<!-- 定义全局结果 -->
<global-results>
    <!-- 在 Action 中，返回 noLogin 时，跳转到 login.jsp -->
    <result name="noLogin">/login.jsp</result>
    <!-- 对应异常 -->
```

```xml
            <result name="exception">/exception.jsp</result>
            <result name="sqlException">/sqlException.jsp</result>
</global-results>
<!-- 定义异常 -->
<global-exception-mappings>
            <exception-mapping result="exception" exception="java.lang.Exception">
</exception-mapping>
            <exception-mapping result="sqlException" exception="java.sql.SQLException">
</exception-mapping>
            </global-exception-mappings>
            …
</package>
```

### 2.4.2 能力目标

掌握 Action 配置文件中常见的结果类型。

### 2.4.3 任务驱动

1) 任务的主要内容

编写两个 Action 类：FirstAction 和 SecondAction。FirstAction 执行成功后，redirect 到 SecondAction，并向 SecondAction 传递两个参数：uname 和 upwd。程序入口是通过 http://localhost:8080/ch2/first.action 访问 FirstAction，请查看控制台输出结果。

2) 任务的代码模板

FirstAction.java 的代码如下：

```java
package action;
import com.opensymphony.xwork2.ActionSupport;
public class FirstAction extends ActionSupport{
    private static final long serialVersionUID = 1L;
    public String execute(){
        System.out.println("这是第一个 Action,成功后 redirect 到第二个 Action");
        return SUCCESS;
    }
}
```

SecondAction.java 的代码如下：

```java
package action;
import com.opensymphony.xwork2.ActionSupport;
public class SecondAction extends ActionSupport{
    private static final long serialVersionUID = 1L;
    private String uname;        //接收传递过来的参数 uname
    private String upwd;         //接收传递过来的参数 upwd
    public String execute(){
        System.out.println("这是第二个 Action,从第一个 Action 传递过来的参数为 uname = " + uname + ",upwd = " + upwd);
        return SUCCESS;
    }
    public String getUname() {
```

```
            return uname;
        }
        public void setUname(String uname) {
            this.uname = uname;
        }
        public String getUpwd() {
            return upwd;
        }
        public void setUpwd(String upwd) {
            this.upwd = upwd;
        }
}
```

Action 配置文件 struts.xml 的代码模板如下：

```
...
<package name = "hisuser" namespace = "/" extends = "struts-default">
    <action name = "first" class = "action.FirstAction">
        <!-- 代码的功能是从 first.action 重定向(redirect)到 second.action,并传递两
        个参数 uname = ${'chenheng'}和 upwd = ${'123456'} -->
        <result type = "redirect">【代码】</result>
    </action>
    <action name = "second" class = "action.SecondAction">
        <result>/index.jsp</result>
    </action>
</package>
...
```

3) 任务小结或知识扩展

① include 的配置。对于庞大而复杂的系统，模块化设计是一种非常好的思想，对于配置文件也是如此。开发者开发 Web 应用时可按照业务功能模块进行划分，然后为各个业务功能模块分别提供一个 xml 配置文件，最后再将这些 xml 配置文件通过 include 元素统一配置在 struts.xml 文件中。Struts 在解析时，会自动加载这些 xml 配置文件。例如，在 struts.xml 中有如下配置：

```
<struts>
    <include file = "conf/struts-user.xml"/>
    <include file = "conf/struts-admin.xml"/>
    <include file = "conf/struts-common.xml"/>
    ...
</struts>
```

上述 struts.xml 配置文件说明了，在与 struts.xml 同目录下有目录 conf，conf 目录下存有 struts-user.xml、struts-admin.xml 和 struts-common.xml 等子配置文件，这些子配置文件的结构应该与 struts.xml 一样，只是分别处理不同的逻辑模块。这对于多人协同开发来说，非常重要。因此从实际开发来说，struts.xml 就如同上面的配置那样，只将其他子配置文件包含进来，然后再定义一些全局共享的配置，struts.xml 只起到一个统领全局配置文件的作用，更具体的配置都应该放在子配置文件中。

② 包的继承。在 Struts 2 的配置文件中，<package>节点提供了类似于 Java package

的概念，具有继承的特征。子<package>节点既可以继承父<package>节点中的配置，也可以覆盖。一般将一些公共的配置放到父<package>节点中配置，例如：拦截器（后面章节讲解）、全局结果等。<package>继承示例代码如下：

```xml
<package name="my-default" extends="struts-default">
    <!-- 拦截器配置 -->
    <interceptors>
        <interceptor name="myinter" class="intercepter.FirstIntercepter"/>
        <interceptor-stack name="myStack">
            <interceptor-ref name="defaultStack"></interceptor-ref>
            <interceptor-ref name="myinter"></interceptor-ref>
        </interceptor-stack>
    </interceptors>
    <global-results>
        <result name="myyy">/index.jsp</result>
        <result name="exception">/exception.jsp</result>
    </global-results>
    <global-exception-mappings>
        <exception-mapping result="exception" exception="java.lang.Exception">
        </exception-mapping>
    </global-exception-mappings>
    ...
</package>
<!-- 包 hello 继承包 my-default -->
<package name="hello" namespace="/hello" extends="my-default">
    <action name="world" class="action.HelloWorldAction">
        <interceptor-ref name="myStack"></interceptor-ref>
        <result name="success">/hello.jsp</result>
        <!-- 从一个包下的 Action,转到另一个包下的 Action,并传参数 -->
        <result name="fail" type="redirectAction">
            <param name="actionName">world1?id=${id}&amp;uname=${uname}</param>
            <param name="namespace">/hello1</param>
        </result>
    </action>
</package>
```

上述配置文件中，包 my-default 继承了包 struts-default，而包 hello 又继承了包 my-default。需要注意的是，Struts 2 的配置文件是顺序解析的，如果子包 hello 定义在父包 my-default 之前，是错误的。

4）任务代码模板的参考答案

【代码】second.action?uname=${'chenheng'}&amp;upwd=${'123456'}

## 2.4.4 实践环节

请将 Struts 2 Web 应用 ch2 的配置文件 struts.xml 拆分成 3 个 xml 文件：myuser.xml、youruser.xml 以及 hisuser.xml，并将 3 个 xml 文件存放在 conf 目录下，最后使用 include 元素将 3 个 xml 文件统一配置在 struts.xml 主配置文件中。

myuser.xml 的内容如下：

```xml
<?xml version="1.0" encoding="UTF-8"?>
<!DOCTYPE struts PUBLIC "-//Apache Software Foundation//DTD Struts Configuration 2.1//EN"
"http://struts.apache.org/dtds/struts-2.1.dtd">
<struts>
    <package name="myuser" namespace="/user" extends="struts-default">
        <action name="login" class="action.UserAction" method="login">
            <result>/loginSuccess.jsp</result>
        </action>
        <action name="logout" class="action.UserAction" method="logout">
            <result>/login.jsp</result>
        </action>
    </package>
</struts>
```

youruser.xml 的内容如下：

```xml
<?xml version="1.0" encoding="UTF-8"?>
<!DOCTYPE struts PUBLIC "-//Apache Software Foundation//DTD Struts Configuration 2.1//EN"
"http://struts.apache.org/dtds/struts-2.1.dtd">
<struts>
    <package name="youruser" namespace="/" extends="struts-default">
        <action name="myLogin" class="action.MyLoginAction">
            <result>/myLoginSuccess.jsp</result>
            <result name="loginFail">/myLoginFail.jsp</result>
        </action>
        <action name="yourLogin" class="action.YourLoginAction">
            <result>/yourLoginSuccess.jsp</result>
            <result name="fail">/yourLoginFail.jsp</result>
        </action>
        <action name="countNumber" class="action.CountNumberAction">
            <result>/showNumber.jsp</result>
        </action>
    </package>
</struts>
```

hisuser.xml 的内容如下：

```xml
<?xml version="1.0" encoding="UTF-8"?>
<!DOCTYPE struts PUBLIC "-//Apache Software Foundation//DTD Struts Configuration 2.1//EN"
"http://struts.apache.org/dtds/struts-2.1.dtd">
<struts>
    <package name="hisuser" namespace="/" extends="struts-default">
        <action name="first" class="action.FirstAction">
            <result type="redirect">second.action?uname=${'chenheng'}&amp;upwd=${'123456'}</result>
        </action>
        <action name="second" class="action.SecondAction">
            <result>/index.jsp</result>
        </action>
    </package>
</struts>
```

## 2.5 本章小结

本章是整个 Struts 2 框架的核心部分。通过本章的学习，务必掌握如何编写 Action 类以及如何配置 Action。

## 习 题 2

1. result 的类型 redirectAction 表示（　　）。
   A. 处理 Action 链，跳转到下一个 Action　　B. 转发到一个 JSP
   C. 重定向到一个 Action　　D. 重定向到一个 JSP
2. result 的类型 redirect 表示（　　）。
   A. 处理 Action 链，跳转到下一个 Action　　B. 转发到一个 JSP
   C. 重定向到一个 Action 或一个 JSP　　D. 重定向到一个 JSP
3. Struts 2 的默认配置文件是（　　）。
   A. xeb.xml　　B. struts.xml　　C. user.xml　　D. server.xml
4. 某 Action 的配置如下：

```
<action name="hello" class="action.HelloAction">
    <result name="error">/error.jsp</result>
</action>
```

当执行该 Action 时会调用对应类中的方法是（　　）。
   A. execute()　　B. doPost()　　C. doGet()　　D. service()
5. 在 Struts 2 的 Web 应用 hello 中 struts.xml 的配置如下：

```
<package name="user" extends="struts-default" namespace="/my">
    <action name="hello" class="action.HelloAction">
        <result>/success.jsp</result>
    </action>
</package>
```

需要访问该 Action 时，输入的 URL 地址是（　　）。
   A. http://localhost:8080/hello/user/hello.action
   B. http://localhost:8080/hello/my/hello.jsp
   C. http://localhost:8080/hello/my/hello.action
   D. http://localhost:8080/hello/user/hello.jsp
6. result 的类型 dispatcher 表示（　　）。
   A. 处理 Action 链，跳转到下一个 Action　　B. 转发到一个 JSP
   C. 重定向到一个 Action　　D. 重定向到一个 JSP
7. 某 Action 的配置如下：

```
<action name="hello" class="action.HelloAction">
    <result>/main.jsp</result>
```

```
<result name = "sussecc">/success.jsp</result>
<result name = "error">/error.jsp</result>
<result name = "input">/login.jsp</result>
</action>
```

调用该 Action 后返回值是"success",则显示的页面是(　　)。

  A. main.jsp  B. success.jsp  C. error.jsp  D. login.jsp

8. 某 Action 的配置如下：

```
<action name = "hello" >
    <result >/success.jsp</result>
</action>
```

访问该 Action 时会调用的类是(　　)。

  A. Action        B. ActionSupport

  C. HelloAction      D. UserAction

9. 某 Action 的配置如下：

```
<action name = "hello" class = "action.HelloAction" method = "update">
    <result >/success.jsp</result>
</action>
```

访问该 Action 时会执行 HelloAction 类中的方法是(　　)。

  A. execute()  B. delete()  C. select()  D. update()

10. Action 执行的时候并不一定要执行 execute 方法,可以在 URL 地址中动态指定,假设使用动态指定调用 Action：useradd 中的 add 方法,则 URL 地址需要写成(　　)。

  A. ＜a href＝"useradd.action"＞添加用户＜/a＞

  B. ＜a href＝"useradd.action? add"＞添加用户＜/a＞

  C. ＜a href＝"useradd.action! add"＞添加用户＜/a＞

  D. ＜a href＝"useradd.action& add"＞添加用户＜/a＞

11. 在 Struts 2 的 Action 类中访问 Servlet API 有哪几种方法？其中与 HttpServlet 耦合的方法是哪几种？

12. 结果类型 redirect 与 redirectAction 的区别是什么？

13. 在 Action 中接收请求参数有哪几种方式？哪种方式比较方便？

# 第 3 章 Struts 2 的类型转换

**主要内容**

(1) 类型转换的意义。
(2) Struts 2 内置的类型转换器。
(3) 自定义类型转换器。

在 MVC 框架中,需要收集用户请求参数,并将请求参数传递给应用的控制器组件。此时存在一个问题,所有的请求参数类型只能是字符串数据类型,但 Java 是强类型语言,所以 MVC 框架必须将这些字符串请求参数转换成相应的数据类型。

Struts 2 不仅提供了强大的类型转换机制,而且开发者还可以方便地开发出自己的类型转换器,完成字符串和各种数据类型之间的转换。这正是学习本章的目的所在。

## 3.1 Struts 2 内置的类型转换器

### 3.1.1 核心知识

**1. 类型转换的意义**

本节以一个简单应用(JSP+Servlet)为示例来介绍类型转换的意义。如图 3.1 所示的添加商品页面,该页面用于收集用户输入的商品信息。商品信息包括商品名称(字符串类型 String)、商品价格(双精度浮点类型 double)、商品数量(整数类型 int)。

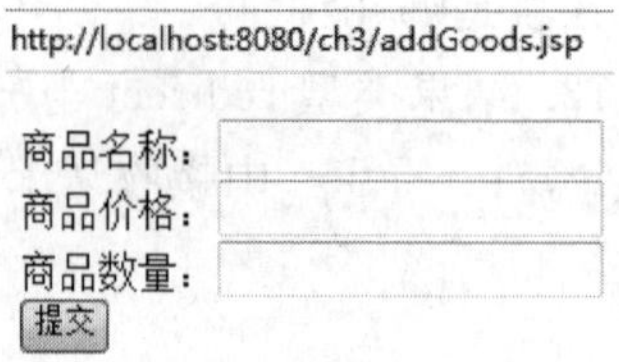

addGoods.jsp 页面的代码如下:

图 3.1 添加商品信息的收集页面

```
<body>
    <form action = "addGoods" method = "post">
        商品名称:<input type = "text" name = "goodsname"/><br>
        商品价格:<input type = "text" name = "goodsprice"/><br>
        商品数量:<input type = "text" name = "goodsnumber"/><br>
        <input type = "submit" value = "提交"/>
    </form>
</body>
```

希望页面收集到的数据提交到 addGoods 的 Servlet(AddGoodsServlet 类),该 Servlet 将这些请求信息封装成一个 Goods 类的值对象。

Goods 类的代码如下:

```java
package model;
public class Goods {
    private String goodsname;
    private double goodsprice;
    private int goodsnumber;
    //无参数的构造方法
    public Goods(){}
    //有参数的构造方法
    public Goods(String goodsname, double goodsprice, int goodsnumber) {
        super();
        this.goodsname = goodsname;
        this.goodsprice = goodsprice;
        this.goodsnumber = goodsnumber;
    }
    //此处省略了 setter 和 getter 方法
    ...
}
```

AddGoodsServlet 类的代码如下:

```java
package servlet;
import java.io.IOException;
import javax.servlet.ServletException;
import javax.servlet.http.HttpServlet;
import javax.servlet.http.HttpServletRequest;
import javax.servlet.http.HttpServletResponse;
import model.Goods;
public class AddGoodsServlet extends HttpServlet {
    public void doGet(HttpServletRequest request, HttpServletResponse response)
        throws ServletException, IOException {
        doPost(request, response);
    }
    public void doPost(HttpServletRequest request, HttpServletResponse response)
        throws ServletException, IOException {
        response.setContentType("text/html;charset=utf-8");
        //设置编码,防止乱码
        request.setCharacterEncoding("utf-8");
        //获取参数值
        String goodsname = request.getParameter("goodsname");
        String goodsprice = request.getParameter("goodsprice");
        String goodsnumber = request.getParameter("goodsnumber");
        //下面进行类型转换
        double newgoodsprice = Double.parseDouble(goodsprice);
        int newgoodsnumber = Integer.parseInt(goodsnumber);
        //将转换后的数据封装成 goods 值对象
```

```
        Goods goods = new Goods(goodsname, newgoodsprice, newgoodsnumber);
        //将 goods 值对象传递给数据访问层,进行添加操作,代码省略
        ...
    }
}
```

对于上面这个应用而言,开发者需要自己在 Servlet 中进行类型转换,并将其封装成值对象。这些类型转换操作全部手工完成,异常烦琐。

对于 MVC 框架而言,它必须将请求参数转换成值对象类里各属性对应的数据类型——这就是类型转换的意义。下面介绍如何使用 Struts 2 的支持来完成类型转换。

### 2. Struts 2 内置的类型转换器

在 Struts 2 框架中,对于常用的数据类型,开发者无须创建自己的类型转换器,因为 Struts 2 有许多内置的类型转换器完成常用的类型转换。Struts 2 提供的内置类型转换器,包括如下几种类型。

（1）boolean 和 Boolean：完成 String 和布尔型之间的转换。

（2）char 和 Character：完成 String 和字符型之间的转换。

（3）int 和 Integer：完成 String 和整型之间的转换。

（4）long 和 Long：完成 String 和长整型之间的转换。

（5）float 和 Float：完成 String 和单精度浮点型之间的转换。

（6）double 和 Double：完成 String 和双精度浮点型之间的转换。

（7）Date：完成 String 和日期类型之间的转换,日期格式为用户请求本地的 SHORT 格式(如：yy-mm-dd)。

（8）数组(arrays)：该类型在数据转换时,必须满足需要转换的数据中每一个元素都能转换成数组的类型。

（9）集合(collections)：在使用集合类型转换器时,如果集合中的数据无法确定,可以先将其封装到一个 String 类型的集合中,然后在用到某个元素时再进行手动转换。

类型转换是在页面与 Action 相互传递数据时发生的。Struts 2 对于基本类型如 int、long、float、double、boolean 以及 char 等(包括 Date),已经做好了基本类型转换。因此如果 Action 中包含基本类型属性(一定要有 getter 和 setter 方法),在页面上只要包含对应此属性的名称,Struts 2 会自动进行类型转换。比如,有这样一个 Action,代码如下：

```
public class TestAction extends ActionSupport{
    //用来封装用户请求中用户名的信息
    private String uname;
    //用来封装用户请求中年龄的信息
    private int uage;
    public String getUname() {
        return uname;
    }
    public void setUname(String uname) {
        this.uname = uname;
    }
    public int getUage() {
```

```java
        return uage;
    }
    public void setUage(int uage) {
        this.uage = uage;
    }
    public String execute(){
        System.out.println(uage);
        return SUCCESS;
    }
}
```

另外,还需要提供一个 JSP 页面,代码如下:

```html
<body>
    <form action = "input.action" method = "post">
    <h3>信息提交</h3>
        用户名:<input type = "text" name = "uname"><br>
        年龄:<input type = "text" name = "uage"><br>
        <input type = "submit" value = "提交">
    </form>
</body>
```

当 JSP 页面的 form 表单提交到 Action 时,Struts 2 使用内置的类型转换器自动将表单中的参数"uage"转换成 int 类型。上述的 Action(TestAction)代码还可以是如下写法:

```java
package action;
import model.Information;
import com.opensymphony.xwork2.ActionSupport;
import com.opensymphony.xwork2.ModelDriven;
public class ReceiveAction extends ActionSupport implements ModelDriven<Information>{
    //定义 info 对象,需要创建该对象,不会主动帮你创建
    private Information info = new Information();
    /**
     * 处理用户请求的方法
     */
    public String execute(){
        System.out.println("uname:" + info.getUname());
        System.out.println("uage:" + info.getUage());
        return SUCCESS;
    }
    @Override
    public Information getModel() {
        return info;
    }
}
```

上面 Action 中的 Model 类 Information 的代码如下:

```java
package model;
public class Information {
    //用来封装用户请求中用户名的信息
    private String uname;
```

```java
        //用来封装用户请求中年龄的信息
        private int uage;
        public String getUname() {
            return uname;
        }
        public void setUname(String uname) {
            this.uname = uname;
        }
        public int getUage() {
            return uage;
        }
        public void setUage(int uage) {
            this.uage = uage;
        }
}
```

### 3.1.2 能力目标

理解类型转换的意义，掌握 Struts 2 内置类型转换器的用法。

### 3.1.3 任务驱动

1) 任务的主要内容

首先，创建一个 JSP 页面 task_3_1.jsp，在 task_3_1.jsp 页面中输入矩形的长与宽（double 类型数据），提交给 Action 类 ComputerAction。其次，在 ComputerAction 类中接收矩形的长与宽，并计算矩形的面积，输出在控制台。

2) 任务的代码模板

task_3_1.jsp 的代码如下：

```jsp
<%@ page language="java" import="java.util.*" pageEncoding="UTF-8"%>
<%
    String path = request.getContextPath();
    String basePath = request.getScheme() + "://" + request.getServerName() + ":" + request.getServerPort() + path + "/";
%>
<!DOCTYPE HTML PUBLIC "-//W3C//DTD HTML 4.01 Transitional//EN">
<html>
  <head>
    <base href="<%=basePath%>">
    <title>My JSP 'task_3_1.jsp' starting page</title>
  </head>
  <body>
    <form action="computer.action" method="post">
        长：<input type="text" name="length"><br>
        宽：<input type="text" name="width"><br>
        <input type="submit" value="提交">
    </form>
  </body>
</html>
```

ComputerAction.java 的代码模板如下：

```java
package action;
import com.opensymphony.xwork2.ActionSupport;
public class ComputerAction extends ActionSupport{
    private static final long serialVersionUID = 1L;
    //代码1 接收矩形长度
    【代码1】
    //代码2 接收矩形宽度
    【代码2】
    public String execute(){
        System.out.println("矩形的面积为：" + length * width);
        return SUCCESS;
    }
    public double getLength() {
        return length;
    }
    public void setLength(double length) {
        this.length = length;
    }
    public double getWidth() {
        return width;
    }
    public void setWidth(double width) {
        this.width = width;
    }
}
```

Action 配置文件内容略。

3）任务小结或知识扩展

Struts 2 内置的类型转换器，使用方便，但需要注意的是页面输入数据类型必须与 Action 接收类型相同或兼容，否则会报"ognl.MethodFailedException"异常。在配置 Action 时，需要给一个名为"input"的 result，才能处理该异常。

4）任务代码模板的参考答案

【代码1】 private double length;

【代码2】 private double width;

### 3.1.4 实践环节

请将本节任务中 Action 类接收请求参数的方式改成 ModelDriven＜T＞方式。

## 3.2 自定义类型转换器

当 Struts 2 内置的类型转换器不能满足需求时，开发者可以开发自己的类型转换器。例如有个应用 ch3 希望使用用户在页面上输入的表单信息来创建商品。当输入"苹果，10.58，200"时，程序自动 new Goods，并将"苹果"值自动赋值给 goodsname 属性，将"10.58"值自动赋值给 goodsprice 属性，将"200"值自动赋值给 goodsnumber 属性。

想实现上述应用功能需要做以下 4 件事。
(1) 需要一个实体类 Goods,封装商品信息。
(2) 创建一个 Action 类。
(3) 创建自定义类型转换器。
(4) 配置类型转换器。
具体步骤如下:
第 1 步,编写实体类 Goods。
创建名为 Goods.java 的类文件,代码如下:

```java
package model;
public class Goods {
    private String goodsname;
    private double goodsprice;
    private int goodsnumber;
    public String getGoodsname() {
        return goodsname;
    }
    public void setGoodsname(String goodsname) {
        this.goodsname = goodsname;
    }
    public double getGoodsprice() {
        return goodsprice;
    }
    public void setGoodsprice(double goodsprice) {
        this.goodsprice = goodsprice;
    }
    public int getGoodsnumber() {
        return goodsnumber;
    }
    public void setGoodsnumber(int goodsnumber) {
        this.goodsnumber = goodsnumber;
    }
}
```

第 2 步,编写 Action 类。

创建名为 GoodsConvertAction.java 的 Action 文件,代码如下:

```java
package action;
import model.Goods;
import com.opensymphony.xwork2.ActionSupport;
public class GoodsConvertAction extends ActionSupport{
    //Struts 2 的转换器会自动将请求过来的值转换成 Goods 类型
    private Goods goods;
    public Goods getGoods() {
        return goods;
    }
    public void setGoods(Goods goods) {
        this.goods = goods;
    }
```

```
        public String execute(){
            return SUCCESS;
        }
    }
```

第 3 步,编写自定义类型转换器类。

编写自定义类型转换器见 3.2.1 节核心知识。

第 4 步,注册自定义类型转换器。

实现了自定义类型转换器之后,将该类型转换器注册在 Web 应用中,Struts 2 框架才可以正常使用该类型转换器。注册自定义类型转换器见 3.2.1 节核心知识。

第 5 步,修改配置文件。

根据业务流程,修改 struts.xml 配置文件,具体代码如下:

```xml
<?xml version="1.0" encoding="UTF-8"?>
<!DOCTYPE struts PUBLIC "-//Apache Software Foundation//DTD Struts Configuration 2.1//EN"
    "http://struts.apache.org/dtds/struts-2.1.dtd">
<struts>
    <package name="converter" namespace="/" extends="struts-default">
        <action name="convert" class="action.GoodsConvertAction">
            <result>/showGoods.jsp</result>
        </action>
    </package>
</struts>
```

第 6 步,新建相关页面。

提供输入和输出的 JSP 页面。注意,与以往不同的是输入页面的 form 表单的 input 元素的 name 属性设置为要转换后的目标类型对象 goods。

输入页面 input.jsp 的代码如下:

```html
<form action="convert.action" method="post">
    请输入商品信息(格式为:苹果,10.58,200):
    <input type="text" name="goods"/><br>
    <input type="submit" value="提交"/>
</form>
```

输出页面并未有任何变化,可以使用 EL 表达式将 goods 对象的属性输出到页面,也可以使用 Struts 2 的标签(后续章节讲解)输出到页面。输出页面 showGoods.jsp 的代码如下:

```html
<body>
    您创建的商品信息如下:<br>
    <!-- 使用 EL 表达式取出 Action 类的属性 goods 的值 -->
    商品名为:${goods.goodsname},
    商品价格为:${goods.goodsprice},
    商品数量为:${goods.goodsnumber}.
</body>
```

程序发布后,打开 IE,输入 http://localhost:8080/ch3/input.jsp,显示的页面如图 3.2 所示。

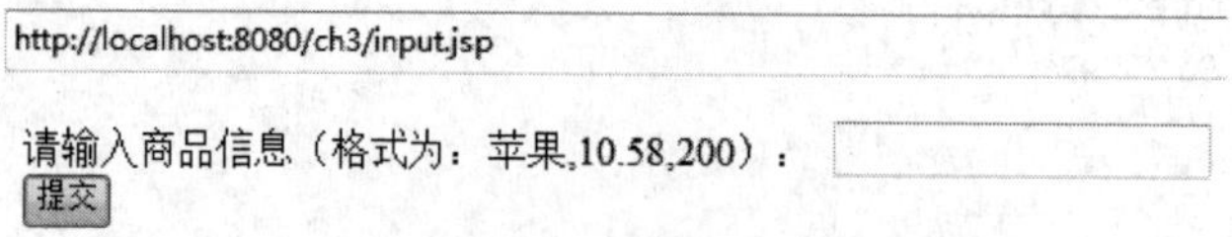

图 3.2　自定义类型转换器示例的首页面

输入"桃子,10.88,800",提交表单查看结果,输出如图 3.3 所示。

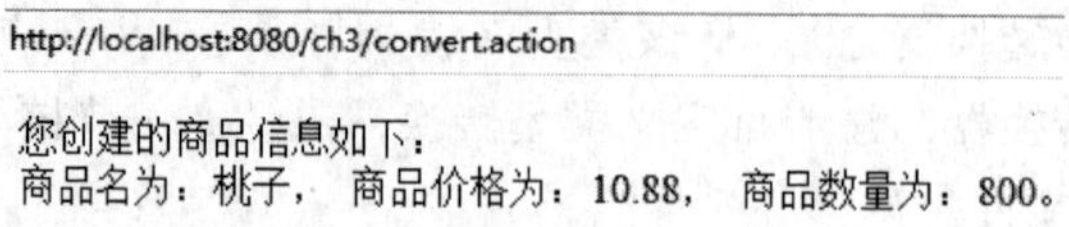

图 3.3　自定义类型转换器示例的结果页面

至此,完成了 Struts 2 的自定义类型转换器。

## 3.2.1　核心知识

### 1. 创建自定义类型转换器

创建自定义类型转换器类,有 3 种方法:实现 ognl.TypeConverter 接口、继承 DefaultTypeConverter 类及继承 StrutsTypeConverter 类。

1) 实现 ognl.TypeConverter 接口

TypeConverter 接口有一个接口方法:

public Object convertValue(Map context, Object target, Member member, String propertyName,
　　　　Object value, Class toType);

实现 ognl.TypeConverter 接口创建自定义类型转换器,必须实现上述接口方法,不过该接口方法过于复杂,所以 OGNL 项目还提供一个 TypeConverter 接口的实现类:DefaultTypeConverter,一般通过继承该类实现自己的类型转换器。

2) 继承 DefaultTypeConverter 类

DefaultTypeConverter 类实现了 TypeConverter 接口,并提供了一个简化的 convertValue 方法。

public Object convertValue(Map context, Object value, Class toType)

convertValue 方法负责完成类型的转换,这种转换是双向的。例如,当需要把字符串转换成 Goods 实例时,是通过该方法实现的;当需要把 Goods 实例转换成字符串时,也是通过该方法实现的。应用 ch3 的自定义类型转换器类 GoodsConverter1.java 的代码如下:

```
package converter;
import java.util.Map;
import model.Goods;
import ognl.DefaultTypeConverter;
//通过继承 DefaultTypeConverter 实现自定义类型转换器
public class GoodsConverter1 extends DefaultTypeConverter{
    //类型转换器必须重写该方法,实现双向转换
    public Object convertValue(Map context, Object value, Class toType) {
```

```java
        //当需要把字符串向 Goods 类型转换时
        if(toType == Goods.class){
            //系统的请求参数是一个字符串数组
            String params[] = (String[])value;
            //创建一个 Goods 实例
            Goods goods = new Goods();
            //只处理请求参数数组的第一个元素(因为页面只有一个请求参数),并以","分隔
            String stringValues[] = params[0].split(",");
            //为 Goods 实例赋值
            goods.setGoodsname(stringValues[0]);
            goods.setGoodsprice(Double.parseDouble(stringValues[1]));
            goods.setGoodsnumber(Integer.parseInt(stringValues[2]));
            return goods;
        }
        //当需要将 Goods 实例向字符串转换时
        else if(toType == String.class){
            Goods goods = (Goods)value;
            return "[" + goods.getGoodsname() + ","
                    + goods.getGoodsprice() + ","
                    + goods.getGoodsnumber() + "]";
        }
        return null;
    }
}
```

3) 继承 StrutsTypeConverter 类

Struts 2 提供了一个 TypeConverter 接口的默认实现类 StrutsTypeConverter。该实现类有两个抽象方法:public Object convertFromString(Map context,String[] values, Class toClass);public String convertToString(Map context,Object obj),在定义类型转换器必须被实现。

convertFromString 方法的功能是将一个或多个字符串值转换为指定的类型。参数 context 是表示 Action 上下文的 Map 对象,参数 values 是要转换的字符串值,参数 toClass 是要转换的目标类型。

convertToString 方法的功能是将指定的对象转换为字符串。参数 context 是表示 Action 上下文的 Map 对象,参数 obj 是要转换的对象。

应用 ch3 的自定义类型转换器类 GoodsConverter2.java 的代码如下:

```java
package converter;
import java.util.Map;
import model.Goods;
import org.apache.struts2.util.StrutsTypeConverter;
public class GoodsConverter2 extends StrutsTypeConverter{
    //将字符串值转换为指定的类型
    @Override
    public Object convertFromString(Map context, String[] values, Class toClass) {
        if(values.length > 0){
            //创建一个 Goods 实例
            Goods goods = new Goods();
```

```
            //只处理请求参数数组的第一个元素(因为页面只有一个请求参数),并以","分隔
            String stringValues[] = values[0].split(",");
            //为 Goods 实例赋值
            goods.setGoodsname(stringValues[0]);
            goods.setGoodsprice(Double.parseDouble(stringValues[1]));
            goods.setGoodsnumber(Integer.parseInt(stringValues[2]));
            return goods;
        }else{
            return null;
        }
    }
    //将指定的对象转换为字符串
    @Override
    public String convertToString(Map context, Object obj) {
        if(obj instanceof Goods){
            Goods goods = (Goods)obj;
            return "[" + goods.getGoodsname() + ","
                    + goods.getGoodsprice() + ","
                    + goods.getGoodsnumber() + "]";
        }
        return null;
    }
}
```

### 2. 注册类型转换器

在 Struts 2 Web 应用中,注册类型转换器有两种常用方式:注册局部类型转换器:仅仅对某个 Action 的属性起作用,注册全局类型转换器:所有 Action 的特定属性都会生效。

1) 注册局部类型转换器

在 Action 所在的包中建立 properties 文件,文件命名格式为 ActionName-conversion.properties。ActionName 是需要转换器生效的 Action 的类名,后面的-conversion.properties 是固定部分。

对于本节中 GoodsConvertAction 类,应该提供的类型转换器注册文件的文件名为:GoodsConvertAction-conversion.properties,该文件是一个典型的 properties 文件,文件由 key-value 对组成。文件内容为:

propertyName = 类型转换器类

ActionName-conversion.properties 文件由多个"propertyName=类型转换器类"项组成,其中 propertyName 是 Action 中需要类型转换器转换的属性名,类型转换器类是开发者实现的类型转换器的全限定类名(需要加包名)。

下面是 GoodsConvertAction-conversion.properties 文件的内容(使用的自定义类型转换器类是本节中的 GoodsConverter1.java):

goods = converter.GoodsConverter1

至此,局部类型转换器注册成功。

2）注册全局类型转换器

假设应用中有多个 Action 都包含了 Goods 类型的属性，如果多次重复注册局部类型转换器，则将是很烦琐的事情。幸运的是 Struts 2 提供了全局类型转换器，它对指定类型的全部属性有效。

注册全局类型转换器应该在 classpath(src)下提供一个 xwork-conversion.properties 文件，其内容由多个"目标类型＝对应类型转换器"项组成，其中"目标类型"指定需要完成类型转换的类（全限定名），"对应类型转换器"指定类型转换的转换器类（全限定名）。例如，指定 model.Goods 类的类型转换器为本节中的 converter.GoodsConverter2，则注册全局类型转换器的注册文件代码如下：

```
model.Goods = converter.GoodsConverter2
```

一旦注册了上面的全局类型转换器，该全局类型转换器就会对所有 Goods 类型属性起作用。

**注意**：如果 Action 的某个属性类型既被注册了局部类型转换器，又被注册了全局类型转换器，则执行局部类型转换器。

### 3.2.2 能力目标

了解自定义类型转换器的创建与注册。

### 3.2.3 任务驱动

1）任务的主要内容

① 编写一个 JSP 页面 myInput.jsp，该页面运行效果如图 3.4 所示。

② 编写实体类 User，包括 uname 和 upass 属性。

③ 编写 Action 类，在 Action 类中，类型转换器会自动将请求过来的值转换成 User 类型。

④ 通过继承 StrutsTypeConverter 类的方式，编写自定义类型转换器类 UserConverter。

⑤ 注册全局类型转换器。

⑥ 配置 Action。

⑦ 编写用户信息输出页面 showUser.jsp，页面效果如图 3.5 所示。

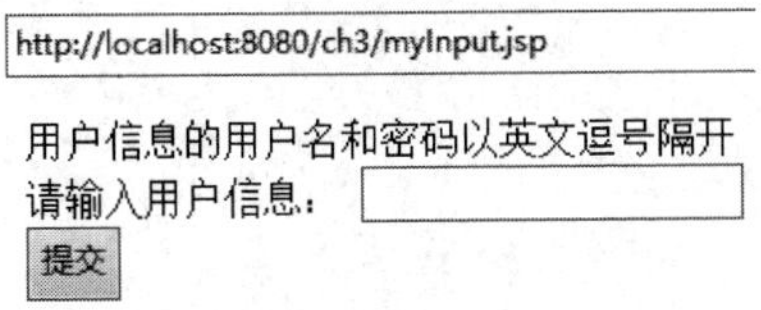

图 3.4 任务首页面

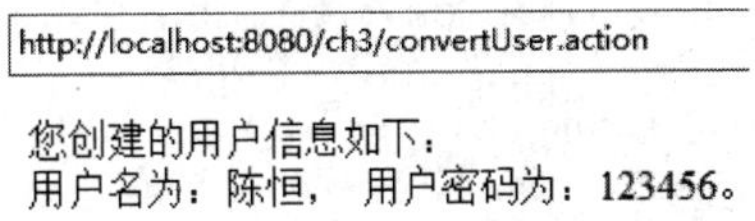

图 3.5 任务结果页面

2）任务的代码模板

myInput.jsp 的代码如下：

```
<%@ page language="java" import="java.util.*" pageEncoding="UTF-8"%>
<%
```

```jsp
String path = request.getContextPath();
String basePath = request.getScheme() + "://" + request.getServerName() + ":" + request.getServerPort() + path + "/";
%>
<!DOCTYPE HTML PUBLIC "-//W3C//DTD HTML 4.01 Transitional//EN">
<html>
  <head>
    <base href="<%=basePath%>">
    <title>My JSP 'myInput.jsp' starting page</title>
  </head>
  <body>
    <form action="convertUser.action" method="post">
        用户信息的用户名和密码以英文逗号隔开<br>
        请输入用户信息:
        <input type="text" name="user"/><br>
        <input type="submit" value="提交"/>
    </form>
  </body>
</html>
```

实体类 User.java 的代码如下:

```java
package model;
public class User {
    private String uname;
    private String upass;
    public String getUname() {
        return uname;
    }
    public void setUname(String uname) {
        this.uname = uname;
    }
    public String getUpass() {
        return upass;
    }
    public void setUpass(String upass) {
        this.upass = upass;
    }
}
```

Action 类 UserConvertAction.java 的代码如下:

```java
package action;
import model.User;
import com.opensymphony.xwork2.ActionSupport;
public class UserConvertAction extends ActionSupport{
    private static final long serialVersionUID = 1L;
    private User user;
    public User getUser() {
        return user;
    }
    public void setUser(User user) {
```

```
            this.user = user;
        }
        public String execute(){
            return SUCCESS;
        }
}
```

自定义类型转换器类 UserConverter.java 的代码模板如下：

```
package converter;
import java.util.Map;
import model.User;
import org.apache.struts2.util.StrutsTypeConverter;
public class UserConverter extends 【代码 1】{
    @Override
    public Object convertFromString(Map context, String[] values, Class toClass) {
        if(values.length > 0){
            User user = new User();
            String stringValues[] = values[0].split(",");
            user.setUname(stringValues[0]);
            user.setUpass(stringValues[1]);
            return user;
        }
        return null;
    }
    @Override
    public String convertToString(Map context, Object obj) {
        if(obj instanceof User){
            User user = (User)obj;
            return "[" + user.getUname() + ","
                    + user.getUpass() + "]";
        }
        return null;
    }
}
```

注册全局类型转换器的代码模板如下：

```
model.User =【代码 2】
```

配置 Action 的代码略。

showUser.jsp 的代码如下：

```
<%@ page language="java" import="java.util.*" pageEncoding="utf-8"%>
<%
    String path = request.getContextPath();
    String basePath = request.getScheme() + "://" + request.getServerName() + ":" + request.getServerPort() + path + "/";
%>
<!DOCTYPE HTML PUBLIC "-//W3C//DTD HTML 4.01 Transitional//EN">
<html>
  <head>
```

```html
<base href="<%=basePath%>">
<title>My JSP 'showUser.jsp' starting page</title>
</head>
<body>
    您创建的用户信息如下：<br>
    用户名为：${User.uname}，
    用户密码为：${User.upass}。
</body>
</html>
```

3) 任务小结或知识扩展

① 数组类型转换器。一直只处理字符串数组的第一个数组元素——请求参数是单个值，而不是一个字符串数组。实际上，请求参数经常是字符串数组的情形，考虑到如图3.6所示的输入页面。

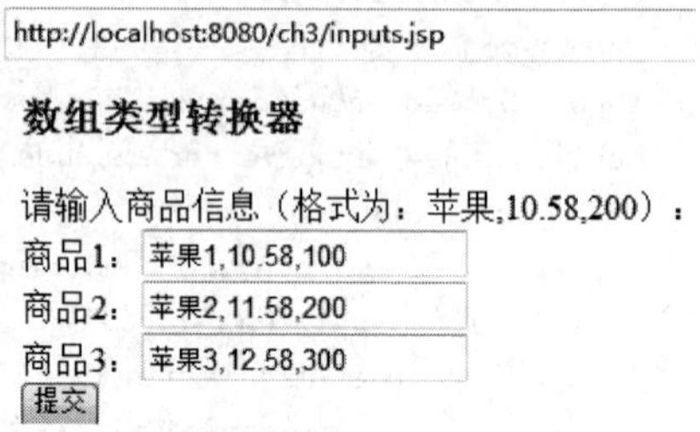

上述页面 inputs.jsp 的代码如下：

```html
<body>
    <h3>数组类型转换器</h3>
    <form action="converts.action" method="post">
    请输入商品信息(格式为：苹果,10.58,200)：<br>
    商品1：<input type="text" name="goods"/><br>
    商品2：<input type="text" name="goods"/><br>
    商品3：<input type="text" name="goods"/><br>
    <input type="submit" value="提交"/>
    </form>
</body>
```

图 3.6 数组类型转换器的首页面

在此页面中包含3个商品信息的请求参数，名称都是 goods。此时 goods 请求参数必须是数组类型或 List 类型。

下面是处理该请求的 Action 类的代码：

```java
package action;
import model.Goods;
import com.opensymphony.xwork2.ActionSupport;
public class GoodsConvertActions extends ActionSupport{
    private Goods[] goods;
    public Goods[] getGoods() {
        return goods;
    }
    public void setGoods(Goods[] goods) {
        this.goods = goods;
    }
    public String execute(){
        return SUCCESS;
    }
}
```

上面的 Action 使用了 Goods[] 数组类型属性来封装 goods 请求参数。下面是类型转换器代码：

```java
package converter;
import java.util.Map;
import model.Goods;
import org.apache.struts2.util.StrutsTypeConverter;
public class GoodsConverter3 extends StrutsTypeConverter{
    //将字符串值转换为指定的类型
    @Override
    public Object convertFromString(Map context, String[] values, Class toClass) {
        if(values.length > 1){
            //创建一个数组Goods实例
            Goods goods[] = new Goods[values.length];
            //遍历数组
            for(int i = 0; i < values.length; i++){
                //将每个数组元素转换成一个Goods实例
                Goods agoods = new Goods();
                String stringValues[] = values[i].split(",");
                //为agoods实例赋值
                agoods.setGoodsname(stringValues[0]);
                agoods.setGoodsprice(Double.parseDouble(stringValues[1]));
                agoods.setGoodsnumber(Integer.parseInt(stringValues[2]));
                goods[i] = agoods;
            }
            return goods;
        }else{
            return null;
        }
    }
    //将指定的对象转换为字符串
    @Override
    public String convertToString(Map context, Object obj) {
        if(obj instanceof Goods[]){
            Goods goods[] = (Goods[])obj;
            String result = "[";
            //遍历数组
            for(Goods agoods : goods){
                result = result + "<" +
                    agoods.getGoodsname() + ","
                    + agoods.getGoodsprice() + ","
                    + agoods.getGoodsnumber() + ">";
            }
            return result + "]";
        }
        return null;
    }
}
```

注册上述类型转换器和配置Action的代码略。

当在如图3.6所示页面的3个文本框内分别输入：苹果1,10.58,100；苹果2,11.58,200；苹果3,12.58,300后,提交请求,将看到如图3.7所示的页面。

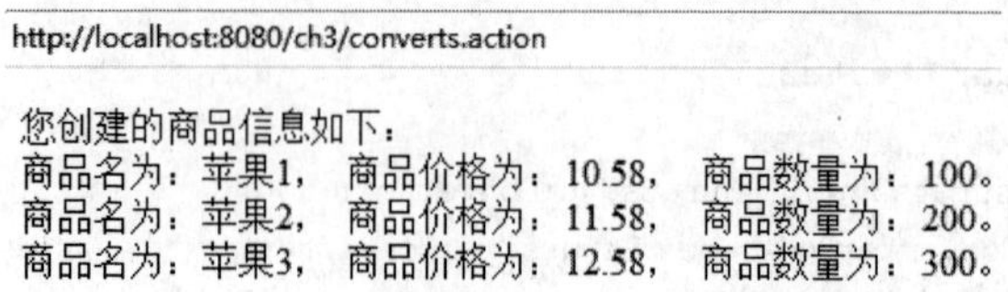

```
http://localhost:8080/ch3/converts.action
```

您创建的商品信息如下:
商品名为: 苹果1, 商品价格为: 10.58, 商品数量为: 100。
商品名为: 苹果2, 商品价格为: 11.58, 商品数量为: 200。
商品名为: 苹果3, 商品价格为: 12.58, 商品数量为: 300。

图 3.7 数组类型转换器的结果页面

上述结果页面 showGoodses.jsp 的代码如下:

```jsp
<%@ page language="java" import="java.util.*" pageEncoding="utf-8"%>
<%
    String path = request.getContextPath();
    String basePath = request.getScheme() + "://" + request.getServerName() + ":" + request.getServerPort() + path + "/";
%>
<%@ taglib prefix="c" uri="http://java.sun.com/jsp/jstl/core" %>
<!DOCTYPE HTML PUBLIC "-//W3C//DTD HTML 4.01 Transitional//EN">
<html>
  <head>
    <base href="<%=basePath%>">
    <title>My JSP 'showGoods.jsp' starting page</title>
  </head>
  <body>
    您创建的商品信息如下:<br>
    <c:forEach var="mf" items="${goods}">
    商品名为: ${mf.goodsname},
    商品价格为: ${mf.goodsprice},
    商品数量为: ${mf.goodsnumber}.<br>
    </c:forEach>
  </body>
</html>
```

② 集合属性类型转换器。如果 goods 属性不是使用字符串数组封装,而是使用集合属性来处理,则修改后的 Action 代码如下:

```java
package action;
import java.util.List;
import model.Goods;
import com.opensymphony.xwork2.ActionSupport;
public class GoodsConvertActionsList extends ActionSupport{
    private List<Goods> goods;
    public List<Goods> getGoods() {
        return goods;
    }
    public void setGoods(List<Goods> goods) {
        this.goods = goods;
    }
    public String execute(){
        return SUCCESS;
    }
}
```

因为 Action 里的 goods 属性是一个 List 类型，应该在类型转换器中也提供对应的转换，因此提供如下的类型转换器：

```java
package converter;
import java.util.ArrayList;
import java.util.List;
import java.util.Map;
import model.Goods;
import org.apache.struts2.util.StrutsTypeConverter;
public class GoodsConverter4 extends StrutsTypeConverter{
    //将字符串值转换为指定的类型
    @Override
    public Object convertFromString(Map context, String[] values, Class toClass) {
        if(values.length > 1){
            //创建一个 List 对象
            List<Goods> result = new ArrayList<Goods>();
            //遍历请求参数数组
            for(int i = 0; i < values.length; i++){
                //将每个数组元素转换成一个 Goods 实例
                Goods agoods = new Goods();
                String stringValues[] = values[i].split(",");
                //为 agoods 实例赋值
                agoods.setGoodsname(stringValues[0]);
                agoods.setGoodsprice(Double.parseDouble(stringValues[1]));
                agoods.setGoodsnumber(Integer.parseInt(stringValues[2]));
                result.add(agoods);
            }
            return result;
        }else{
            return null;
        }
    }
    //将指定的对象转换为字符串
    @Override
    public String convertToString(Map context, Object obj) {
        if(obj instanceof Goods[]){
            List<Goods> goods = (List<Goods>)obj;
            String result = "[";
            //遍历数组
            for(Goods agoods : goods){
                result = result + "<" +
                        agoods.getGoodsname() + ","
                        + agoods.getGoodsprice() + ","
                        + agoods.getGoodsnumber() + ">";
            }
            return result + "]";
        }
        return null;
    }
}
```

修改后的代码只需要将对应的配置文件做简单修改即可,JSP 页面不需要做任何修改。实际上,List 对象和数组几乎可以互换使用。即使 Action 类里使用 List 对象来封装请求参数的多个值,类型转换器也可以将字符串数组转换成 Goods 数组(不需要修改类型转换器类),这是因为 Struts 2 默认支持数组和 List 之间的转换。

4) 任务代码模板的参考答案

【代码 1】StrutsTypeConverter

【代码 2】converter.UserConverter

### 3.2.4 实践环节

将任务中自定义类型转换器类 UserConverter,修改为"继承 DefaultTypeConverter"的方式。

## 3.3 本章小结

本章重点讲解了自定义类型转换器的实现和注册。但在实际应用中,开发者很少自定义类型转换器,一般都是使用 Struts 2 内置的转换器。因为,Struts 2 提供了常用类型的转换功能。

## 习 题 3

1. 在 MVC 框架中,为什么要进行类型转换?
2. Struts 2 提供了哪些内置的类型转换器?
3. 在 Struts 2 框架中,如何自定义类型转换器类,又如何注册类型转换器?

# Struts 2 的拦截器

**主要内容**

(1) 自定义拦截器。
(2) 内置拦截器。
(3) 拦截器的配置。

拦截器(Interceptor)是 Struts 2 框架的核心组成部分。在 Struts 2 文档中对拦截器的解释为拦截器是动态拦截 Action 调用的对象。它提供了一种机制,使开发者可以定义一个特定的功能模块,这个模块可以在 Action 执行之前或者之后运行,也可以在一个 Action 执行之前阻止 Action 执行。

在 Struts 2 框架中,当需要使用某个拦截器时,只需要在配置文件中进行相关的配置即可;如果不需要使用某个拦截器,只需要在配置文件中取消该拦截器的配置即可。在实际应用中,不管是否应用某个拦截器,对整个 Struts 2 框架不会产生任何影响。这是一种可插拔式的设计,具有非常好的可扩展性。

## 4.1 拦截器的定义与配置

### 4.1.1 核心知识

**1. 拦截器的原理**

Struts 2 的拦截器实现相对简单。当请求到达 Struts 2 的 StrutsPrepareAndExecuteFilter 时,Struts 2 会查找配置文件,并根据其配置实例化对应的拦截器对象,然后串成一个拦截器栈(又称拦截器链),最后一个一个地调用栈中的拦截器。

拦截器围绕着 Action 和 Result 的执行而执行,其工作方式如图 4.1 所示。

从图 4.1 可以看到,在 Action 和 Result 执行之前,为 Action 配置的拦截器将首先被执行,在 Action 和 Result 执行之后,拦截器将重新获得控制权,然后按照与先前调用相反的顺序依次执行。在整个执行过程中,任何一个拦截器都可以选择直接返回,从而终止余下的拦截器、Action 和 Result 的执行。例如:当一个未授权的用户访问受保护的资源时,执行身份验证的拦截器可以直接返回。

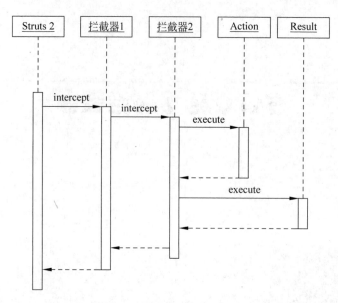

图 4.1　拦截器的工作方式

### 2. Struts 2 内置的拦截器

Struts 2 框架利用大量内置的拦截器，完成了框架内的大部分操作，例如请求参数解析、文件的上传和下载、国际化、类型转换器和数据校验等。这些内置拦截器以 name-class 对的形式配置在 struts-default.xml 文件中，其中 name 是拦截器的名字，就是以后使用该拦截器的唯一标识；class 则指定了该拦截器的实现类。在配置文件中，如果定义的 package 继承了 Struts 2 的 struts-default 包，则可以自由使用 Struts 2 内置的拦截器，否则必须自己定义这些拦截器。

下面是 Struts 2 内置拦截器的简单介绍。

（1）alias：实现在不同请求中相似参数别名的转换。

（2）autowiring：这是个自动装配的拦截器，主要用于当 Struts 2 和 Spring 整合时，Struts 2 可以使用自动装配的方式来访问 Spring 容器中的 Bean。

（3）chain：构建一个 Action 链，使当前 Action 可以访问前一个 Action 的属性，一般和 < result type＝"chain"…/>一起使用。

（4）conversionError：这是一个负责处理类型转换错误的拦截器，它负责将类型转换错误从 ActionContext 中取出，并转换成 Action 的 FieldError 错误。

（5）createSession：该拦截器负责创建一个 HttpSession 对象，主要用于那些需要有 HttpSession 对象才能正常工作的拦截器中。

（6）debugging：当使用 Struts 2 的开发模式时，这个拦截器会提供更多的调试信息。

（7）execAndWait：后台执行 Action，负责将等待画面发送给用户。

（8）exception：该拦截器负责处理异常，它将异常映射为结果。

（9）fileUpload：这个拦截器主要用于文件上传，它负责解析表单中文件域的内容。

（10）i18n：这是支持国际化的拦截器，它负责把所选的语言、区域放入用户 Session 中。

（11）logger：这是一个负责日志记录的拦截器，主要是输出 Action 的名字。

(12) model-driven：这是一个用于模型驱动的拦截器，当某个 Action 类实现了 ModelDriven 接口时，它负责把 getModel()方法的结果堆入 ValueStack 中。

(13) scoped-model-driven：如果一个 Action 实现了一个 ScopedModelDriven 接口，该拦截器负责从指定生存范围中找出指定的 Model，并将通过 setModel 方法将该 Model 传给 Action 实例。

(14) params：这是最基本的一个拦截器，它负责解析 HTTP 请求中的参数，并将参数值设置成 Action 对应的属性值。

(15) prepare：如果 Action 实现了 Preparable 接口，将调用该拦截器的 prepare()方法。

(16) static-params：该拦截器负责将配置文件中<action>标签下<param>标签中的参数传入 Action。

(17) scope：这是范围转换拦截器，它可以将 Action 状态信息保存到 HttpSession 范围，或者保存到 ServletContext 范围内。

(18) servlet-config：如果某个 Action 需要直接访问 Servlet API，就是通过该拦截器实现的。

(19) timer：该拦截器负责输出 Ation 的执行时间，它在分析该 Action 的性能瓶颈时比较有用。

(20) token：该拦截器主要用于阻止重复提交，它检查传到 Action 中的 token，从而防止多次提交。

(21) token-session：该拦截器的作用与前一个基本类似，只是它把 token 保存在 HttpSession 中。

(22) validation：通过执行在 xxxAction-validation.xml 中定义的校验器，从而完成数据校验。

(23) workflow：该拦截器负责调用 Action 类中的 validate 方法，如果校验失败，则返回 input 的逻辑视图。

一般情况下，开发者无须手动控制这些内置的拦截器，因为 struts-default.xml 文件中已经配置了这些拦截器，只需要定义的包继承 struts-default 包，就可以直接使用这些拦截器。

3. 拦截器的配置

Struts 2 拦截器由 struts-default.xml、struts.xml 等配置文件进行管理，开发者开发自己的拦截器时，需要在 struts.xml 文件中进行配置，然后才能使用自己的拦截器。

1) 配置拦截器

拦截器的配置是在 struts.xml 中完成的，定义一个拦截器使用<interceptor.../>元素，其格式如下：

<interceptor name = "拦截器名" class = "拦截器实现类"></interceptor>

一般情况下，上述这种格式就可完成拦截器的配置。但有的时候，如果需要在配置拦截器时为其传入拦截器参数，只要在<interceptor...>与</interceptor>之间配置<param.../>子标签即可传入相应的参数，其格式如下：

<interceptor name = "拦截器名" class = "拦截器实现类">
    <param name = "参数名">参数值</param>

```
...<!-- 如果需要传入多个参数,可以一并设置 -->
</interceptor>
```

在 struts.xml 中可以配置多个拦截器,它们被包在< interceptors ></interceptors >之间,例如下面的配置:

```
<?xml version = "1.0" encoding = "UTF - 8" ?>
    <! DOCTYPE struts PUBLIC " - //Apache Software Foundation//DTD Struts Configuration 2.0//EN"
    "http://struts.apache.org/dtds/struts - 2.1.dtd">
    < struts >
      < package name = "default" extends = "struts - default">
        < interceptors >
          < interceptor name = "拦截器名 1" class = "拦截器类 1"></interceptor >
          < interceptor name = "拦截器名 2" class = "拦截器类 2"></interceptor >
          …
          < interceptor name = "拦截器名 n" class = "拦截器类 n"></interceptor >
        </interceptors >
        …<!-- action 配置 -->
      </package >
    </struts >
```

拦截器是配置在包下的。在包下配置了一系列的拦截器,但仅仅是配置在该包下,并没有得到应用。如果要应用这些拦截器,就需要在< action >配置中引用这些拦截器,格式如下:

```
< interceptor - ref name = "拦截器名" ></interceptor - ref >
```

在< action >配置中引用拦截器示例如下:

```
< action name = "Action 名" class = "Action 类">
    < interceptor - ref name = "拦截器 1"></interceptor - ref >
    < interceptor - ref name = "拦截器 2"></interceptor - ref >
    < interceptor - ref name = "defaultStack"></interceptor - ref >
</action >
```

2) 拦截器栈

如果需要在 Action 执行前同时做登录检查、安全检查和记录日志等操作,则可以将这 3 个动作对应的拦截器配置到< action >标签下。但是,这样做不仅麻烦,而且不利后期的维护,此时就需要拦截器栈。所谓拦截器栈就是将一些拦截器组合起来进行统一管理,格式如下:

```
< interceptor - stack name = " … ">
  < interceptor - ref name = "拦截器 1"></interceptor - ref >
  < interceptor - ref name = "拦截器 2"></interceptor - ref >
  …
  < interceptor - ref name = "defaultStack"></interceptor - ref >
</interceptor - stack >
```

拦截器栈配置完成后,就可以在< action >中对其引用,一个 Action 引用拦截器栈的格式如下:

```xml
<interceptors>
    <interceptor-stack name="myStack">
        <interceptor-ref name="拦截器1"></interceptor-ref>
        <interceptor-ref name="拦截器2"></interceptor-ref>
        …
        <interceptor-ref name="defaultStack"></interceptor-ref>
    </interceptor-stack>
</interceptors>
…
<action name="Action名" class="Action类">
    <interceptor-ref name="myStack"></interceptor-ref>
</action>
…
```

**注意**：一旦继承了 struts-default 包（package），所有 Action 都会调用默认拦截器栈（defaultStack）。但是当在 Action 配置中加入"<interceptor-ref name="拦截器名"></interceptor-ref>"元素后，则会覆盖 defaultStack，所以在 Action 配置中写拦截器引用时，需要显示引用 defaultStack（最好先引用自定义拦截器，再引用 defaultStack）。

**4. 自定义拦截器**

虽然，Struts 2 为开发者提供了许多内置的拦截器，但是这并不意味开发者失去创建自定义拦截器的能力，恰恰相反，在 Struts 2 框架中自定义拦截器是相当容易的一件事。

实现自定义拦截器的方法有如下 3 种。

（1）实现 Interceptor 接口。

（2）继承 AbstractInterceptor 类。

（3）继承 MethodFilterInterceptor 类。

在编写拦截器类的时候需要注意，拦截器是无状态的，换句话说，在拦截器类中不应该有实例变量。这是因为 Struts 2 对每一个 Action 的请求，使用的是同一个拦截器实例来拦截调用，如果拦截器有状态，在多个线程（客户端的每个请求将由服务器端的一个线程来服务）同时访问一个拦截器实例的情况下，拦截器的状态将不可预测。

**1）实现 Interceptor 接口**

Struts 2 提供了 com.opensymphony.xwork2.interceptor.Interceptor 接口，开发者只要实现该接口就可完成自定义拦截器类的编写。该接口的代码如下：

```java
public interface Interceptor extends Serializable{
    //初始化拦截器回调方法
    void init();
    //拦截器实现拦截的逻辑方法
    String intercept(ActionInvocation invocation) throws Exception;
    //销毁拦截器之前的回调方法
    void destroy();
}
```

通过上面接口的定义可以看出，该接口中有如下 3 种方法。

init()：该方法在拦截器被实例化之后、拦截器执行之前调用。该方法只被执行一次，主要用于初始化资源。

intercept(ActionInvocation invocation)：该方法用于实现拦截的动作。该方法有个参数，用该参数调用 invoke()方法，将控制权交给下一个拦截器，或者交给 Action 类的方法。

destroy()：该方法与 init()方法对应，拦截器实例被销毁之前调用。用于销毁在 init()方法中打开的资源。

下面通过一个应用 ch4 来讲解自定义拦截器的实现步骤。

① 定义 Action 类。为了测试自定义拦截器，首先在应用 ch4 中定义一个 Action 类 TimerTestAction，具体代码如下：

```
package action;
import com.opensymphony.xwork2.ActionSupport;
public class TimerTestAction extends ActionSupport{
    public String execute(){
        try {
            //暂停 1000 毫秒
            Thread.sleep(1000);
        } catch (InterruptedException e) {
            e.printStackTrace();
        }
        return SUCCESS;
    }
}
```

② 通过实现 Interceptor 接口，创建自定义拦截器类。在应用 ch4 中通过实现 Interceptor 接口，创建自定义拦截器类 TimerInterceptor，在该拦截器中计算 Action 执行的时间并输出到控制台，具体代码如下：

```
package interceptors;
import com.opensymphony.xwork2.ActionInvocation;
import com.opensymphony.xwork2.interceptor.Interceptor;
public class TimerInterceptor implements Interceptor{
    @Override
    public void destroy() {
    }
    @Override
    public void init() {
    }
    @Override
    public String intercept(ActionInvocation invocation) throws Exception {
        long startTime = System.currentTimeMillis();
        String result = invocation.invoke();
        long executionTime = System.currentTimeMillis() - startTime;
        System.out.println("Action 的执行花费的毫秒数是：" + executionTime);
        return result;
    }
}
```

③ 编写配置文件 struts.xml。在配置文件中，定义拦截器并在配置 Action 时使用拦截器，具体代码如下：

```xml
<?xml version="1.0" encoding="UTF-8"?>
<!DOCTYPE struts PUBLIC "-//Apache Software Foundation//DTD Struts Configuration 2.1//EN"
"http://struts.apache.org/dtds/struts-2.1.dtd">
<struts>
    <package name="ch4" namespace="/" extends="struts-default">
        <interceptors>
            <!-- 定义拦截器 timer -->
            <interceptor name="timer" class="interceptors.TimerInterceptor">
            </interceptor>
        </interceptors>

        <action name="test" class="action.TimerTestAction">
            <!-- 使用拦截器 timer -->
            <interceptor-ref name="timer"></interceptor-ref>
            <!-- 使用内置拦截器 -->
            <interceptor-ref name="defaultStack"></interceptor-ref>
            <result>/index.jsp</result>
        </action>
    </package>
</struts>
```

④ 运行 Action，测试拦截器。在浏览器地址栏中输入：http://localhost:8080/ch4/test.action，运行 Action 并测试拦截器。

至此，通过上述 4 个步骤，完成了自定义拦截器的创建，并通过运行 Action 测试拦截器。

2）继承 AbstractInterceptor 类

除了 Interceptor 接口之外，Struts 2 框架还提供了 AbstractInterceptor 类，该类实现了 Interceptor 接口，并提供了 init()方法和 destroy()方法的空实现。在一般的拦截器实现中，都会继承该类，因为一般的拦截器是不需要打开资源的，故无须实现这两种方法，继承该类会更简洁方便。

通过继承 AbstractInterceptor 类的方式实现 TimerInterceptor.java 的代码，具体如下：

```java
package interceptors;
import com.opensymphony.xwork2.ActionInvocation;
import com.opensymphony.xwork2.interceptor.AbstractInterceptor;
public class ExtendsAbstractInterceptor extends AbstractInterceptor{
    @Override
    public String intercept(ActionInvocation invocation) throws Exception {
        long startTime = System.currentTimeMillis();
        String result = invocation.invoke();
        long executionTime = System.currentTimeMillis() - startTime;
        System.out.println("Action 的执行花费的毫秒数是：" + executionTime);
        return result;
    }
}
```

3）继承 MethodFilterInterceptor 类

Struts 2 框架还提供了 com.opensymphony.xwork2.interceptor.MethodFilterInterceptor 抽

象类,该类继承了 AbstractInterceptor 类,这个拦截器可以指定要拦截或排除 Action 的方法列表。通常情况下,拦截器将拦截 Action 的所有方法调用,但在某些应用场景中,对某些方法的拦截将会出现一些问题。例如:对表单字段进行验证的拦截器,当通过 doDefault() 方法输出表单时,该方法不应该被拦截,因为此时表单字段都没有数据。

MethodFilterInterceptor 类的部分代码如下:

```java
public abstract class MethodFilterInterceptor extends AbstractInterceptor{
    protected Set<String> excludeMethods = Collections.emptySet();
    protected Set<String> includeMethods = Collections.emptySet();
    …
    protected abstract String doIntercept(ActionInvocation invocation) throws Exception;
    …
}
```

在编写自定义拦截器类时,只需要继承 MethodFilterInterceptor 类,重写 doIntercept 方法即可。在配置拦截器时通过参数 excludeMethods 和 includeMethods 来选择排除或拦截的方法,配置示例如下:

```xml
<interceptors>
    <!-- 定义拦截器 methodFilter -->
    <interceptor name="methodFilter" class="interceptors.ExtendsMethodFilterInterceptor">
    </interceptor>
    <!-- 定义拦截器栈 myStack -->
    <interceptor-stack name="myStack">
        <!-- 使用拦截器 methodFilter -->
        <interceptor-ref name="methodFilter">
            <!-- 通过参数 excludeMethods 指定方法 excludeMethod1 和 excludeMethod2 不需要被拦截,多个方法之间以逗号(,)分隔 -->
            <param name="excludeMethods">excludeMethod1,excludeMethod2</param>
            <!-- 通过参数 includeMethods 指定方法 includeMethod 和 execute 需要被拦截,多个方法之间以逗号(,)分隔 -->
            <param name="includeMethods">includeMethod,execute</param>
        </interceptor-ref>
        <!-- 使用内置拦截器 -->
        <interceptor-ref name="defaultStack"></interceptor-ref>
    </interceptor-stack>
</interceptors>
```

上述配置中拦截器类 ExtendsMethodFilterInterceptor.java 的代码如下:

```java
package interceptors;
import com.opensymphony.xwork2.ActionInvocation;
import com.opensymphony.xwork2.interceptor.MethodFilterInterceptor;
public class ExtendsMethodFilterInterceptor extends MethodFilterInterceptor{
    @Override
    protected String doIntercept(ActionInvocation arg0) throws Exception {
        System.out.println("执行拦截器中动作");
        return arg0.invoke();
    }
}
```

测试上述拦截器 Action 类的代码如下：

```java
package action;
import com.opensymphony.xwork2.ActionSupport;
public class FilterAction extends ActionSupport{
    public String excludeMethod1(){
        System.out.println("不需要拦截的方法 1");
        return SUCCESS;
    }
    public String excludeMethod2(){
        System.out.println("不需要拦截的方法 2");
        return SUCCESS;
    }
    public String includeMethod(){
        System.out.println("需要拦截的方法 includeMethod");
        return SUCCESS;
    }
    public String execute(){
        System.out.println("需要拦截的方法 execute");
        return SUCCESS;
    }
}
```

配置上述 Action 的代码如下：

```xml
<action name="methodIntercept" class="action.FilterAction" method="excludeMethod1">
    <interceptor-ref name="myStack"></interceptor-ref>
    <result>/index.jsp</result>
</action>
```

在地址栏中输入：http://localhost:8080/ch4/methodIntercept.action，运行 Action 后，在控制台中只输出"不需要拦截的方法 1"，而没有输出"执行拦截器中动作"。这说明，执行 Action 中 excludeMethod1()方法时，没有执行拦截器，即 excludeMethod1()方法不需要被拦截。

### 4.1.2 能力目标

理解拦截器的原理，掌握拦截器的自定义与配置。

### 4.1.3 任务驱动

1) 任务的主要内容

首先，定义两个拦截器类：FirstInterceptor 和 SecondInterceptor，FirstInterceptor 拦截器是通过实现 Interceptor 接口编写的，而 SecondInterceptor 是通过继承 AbstractInterceptor 抽象类编写的。

其次，定义一个 Action 类 TestAction，在执行 Action 之前，先执行 FirstInterceptor 和 SecondInterceptor 拦截器。在配置 Action 时，要求使用拦截器栈将两个拦截器与 Action 关联。

## 2) 任务的代码模板

FirstInterceptor.java 的代码模板如下:

```java
package interceptors;
import com.opensymphony.xwork2.ActionInvocation;
import com.opensymphony.xwork2.interceptor.Interceptor;
public class FirstInterceptor implements 【代码1】{
    private static final long serialVersionUID = 1L;
    @Override
    public void destroy() {

    }
    @Override
    public void init() {

    }
    @Override
    public String intercept(ActionInvocation arg0) throws Exception {
        System.out.println("我是第一个拦截器.");
        return arg0.invoke();
    }
}
```

SecondInterceptor.java 的代码模板如下:

```java
package interceptors;
import com.opensymphony.xwork2.ActionInvocation;
import com.opensymphony.xwork2.interceptor.AbstractInterceptor;
public class SecondInterceptor extends 【代码2】{
    private static final long serialVersionUID = 1L;
    @Override
    public String intercept(ActionInvocation arg0) throws Exception {
        System.out.println("我是第二个拦截器.");
        return arg0.invoke();
    }
}
```

Action 类 TestAction.java 的代码如下:

```java
package action;
import com.opensymphony.xwork2.ActionSupport;
public class TestAction extends ActionSupport{
    private static final long serialVersionUID = 1L;
    public String execute(){
        return SUCCESS;
    }
}
```

配置文件的代码模板如下:

```xml
<?xml version="1.0" encoding="UTF-8"?>
<!DOCTYPE struts PUBLIC "-//Apache Software Foundation//DTD Struts Configuration 2.1//EN"
```

```xml
"http://struts.apache.org/dtds/struts-2.1.dtd">
<struts>
    <package name="ch4" namespace="/" extends="struts-default">
        <interceptors>
            ...
            <interceptor name="first" class="interceptors.FirstInterceptor"></interceptor>
            <interceptor name="second" class="interceptors.SecondInterceptor"></interceptor>
            <interceptor-stack name="yourStack">
                <!-- 代码 3 引用 first 拦截器 -->
                 【代码 3】
                <!-- 代码 4 引用 second 拦截器 -->
                 【代码 4】
                <!-- 使用内置拦截器 -->
                <interceptor-ref name="defaultStack"></interceptor-ref>
            </interceptor-stack>
        </interceptors>
        <action name="myTest" class="action.TestAction">
            <!-- 代码 5 引用拦截器栈 yourStack -->
             【代码 5】
            <result>/index.jsp</result>
        </action>
        ...
    </package>
</struts>
```

通过 http://localhost:8080/ch4/myTest.action 测试该任务的拦截器。

3) 任务小结或知识扩展

(1) 拦截器与过滤器的区别具体如下。

① 拦截器不依赖于 Servlet 容器,而过滤器依赖于 Servlet 容器。

② 拦截器只能对 Action 请求起作用,而过滤器则可以对几乎所有的请求起作用。

③ 拦截器可以访问 Action 上下文、值栈里的对象,而过滤器不能。

④ 在 Action 的生命周期中,拦截器可以多次被调用,而过滤器只能在容器初始化时(如 Tomcat 服务器启动)被调用一次。

⑤ 过滤器配置在 web.xml 文件中,而拦截器配置在 struts.xml 文件中。

(2) 默认拦截器。当配置一个包时,可以使用"<default-interceptor-ref name="拦截器名"/>"元素为该包下的所有 Action 配置默认的拦截器。一旦为某个包指定了默认的拦截器,如果该包中的 Action 没有显式引用拦截器,则默认的拦截器将会起作用。但是需要注意的是如果一旦为该包中的 Action 显示引用了某个拦截器,则默认的拦截器就不会起作用,如果该 Action 需要使用该默认拦截器,则必须手动配置该拦截器的引用。

每个<package.../>元素只能有一个<default-interceptor-ref name="拦截器名"/>子元素,即每个包只能指定一个默认的拦截器。

下面是配置默认拦截器的示例:

```xml
<package name="包名">
    <!-- 所有拦截器及拦截器栈都配置在该元素下 -->
    <interceptors>
```

```xml
            <!-- 定义拦截器 -->
            <interceptor name = "拦截器名" class = "拦截器类"/>
             …
            <!-- 定义拦截器栈 -->
            <interceptor-stack name = "拦截器栈名"/>
        </interceptors>
        <!-- 配置该包下的默认拦截器——既可以是拦截器,也可以是拦截器栈 -->
        <default-interceptor-ref name = "拦截器名或拦截器栈名">
            <param name = "参数名">参数值</param>
             …
        </default-interceptor-ref>
        <!-- 配置多个 Action -->
        <action .../>
         …
        </action>
</package>
```

4) 任务代码模板的参考答案

【代码 1】`Interceptor`

【代码 2】`AbstractInterceptor`

【代码 3】`<interceptor-ref name = "first"></interceptor-ref>`

【代码 4】`<interceptor-ref name = "second"></interceptor-ref>`

【代码 5】`<interceptor-ref name = "yourStack"></interceptor-ref>`

### 4.1.4 实践环节

创建一个 Web 应用 project414practice,具体要求如下。

(1) 编写一个 input.jsp 页面,页面运行效果如图 4.2 所示。

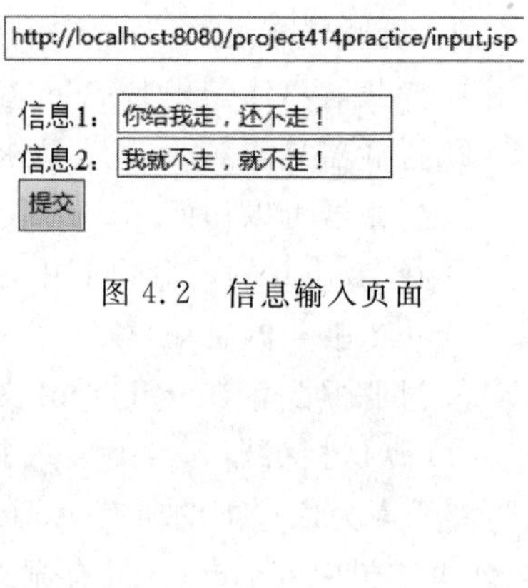

(2) 编写一个自定义拦截器实现对非法文字的拦截,该拦截器类的具体代码如下:

图 4.2 信息输入页面

```java
package interceptor;
import java.util.Iterator;
import java.util.Map;
import java.util.Map.Entry;
import com.opensymphony.xwork2.ActionContext;
import com.opensymphony.xwork2.ActionInvocation;
import com.opensymphony.xwork2.interceptor.AbstractInterceptor;
import com.opensymphony.xwork2.util.ValueStack;
public class InformationFilter extends AbstractInterceptor {
    @Override
    public String intercept(ActionInvocation invocation) throws Exception {
        // 通过核心调度器 invocation 来获得调度的 Action 上下文
        ActionContext actionContext = invocation.getInvocationContext();
        // 获取 Action 上下文的值栈
```

```java
ValueStack stack = actionContext.getValueStack();
// 获取上下文的请求参数
Map valueTreeMap = actionContext.getParameters();
// 获得请求参数集合的迭代器
Iterator iterator = valueTreeMap.entrySet().iterator();
// 遍历请求参数
while (iterator.hasNext()) {
    // 获得迭代的键值对
    Entry entry = (Entry) iterator.next();
    // 获得键值对中的键值
    String key = (String) entry.getKey();
    // 原请求参数,因为有可能一键对多值所以这里用的 String[]
    String[] oldValues = null;
    // 对参数值转换成 String 类型的
    if (entry.getValue() instanceof String) {
        oldValues = new String[] { entry.getValue().toString() };
    } else {
        oldValues = (String[]) entry.getValue();
    }
    // 处理后的请求参数
    String newValueStr = null;
    // 对请求参数过滤处理
    if (oldValues.length > 1) {            //多个请求参数
        newValueStr = "{";
        for (int i = 0; i < oldValues.length; i++) {
            // 替换掉非法参数,这里只替换掉了"走",如有其他需求,
            //可以专门写一个处理字符的类
            newValueStr += oldValues[i].toString().replaceAll("走", "*");
            if (i != oldValues.length - 1) {
                newValueStr += ",";
            }
        }
        newValueStr += "}";
    } else if (oldValues.length == 1) {   //一个请求参数
        newValueStr = oldValues[0].toString().replaceAll("走", "*");
    } else {
        newValueStr = null;
    }
    // 处理后的请求参数加入值栈中
    stack.setValue(key, newValueStr);
}
// 调用下一个拦截器,如果拦截器不存在,则执行 Action
return invocation.invoke();
}
}
```

(3) 编写一个 Action 类,负责获得 input.jsp 页面的信息,并跳转到 showInformation.jsp 页面,该页面显示过滤后的信息,页面运行效果如图 4.3 所示。

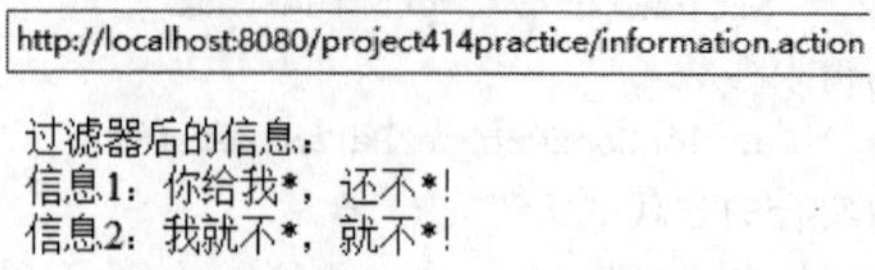

```
http://localhost:8080/project414practice/information.action
```

过滤器后的信息：
信息1：你给我*，还不*！
信息2：我就不*，就不*！

图4.3 过滤后信息显示页面

（4）配置拦截器和Action，并运行input.jsp页面，测试该应用。

## 4.2 使用自定义拦截器完成权限验证

大部分Web应用都涉及权限控制，当使用者需要请求执行某个操作时，Web应用需要先检查使用者是否登录，以及是否有足够的权限来执行该操作。

本节通过Web应用loginInterceptor讲解权限验证的实现，该应用要求用户登录，才可以查看系统中某个视图资源；否则，系统直接转入登录页面。对于上面的需求，可以在每个Action的执行实际处理逻辑之前，先执行权限检查逻辑，但这种做法不利于代码复用。因为大部分Action里的权限检查代码都大同小异，故将这些权限检查的逻辑实现放在拦截器中将会更加合适。

本应用中权限拦截器类的代码如下：

```java
package interceptor;
import java.util.Map;
import com.opensymphony.xwork2.Action;
import com.opensymphony.xwork2.ActionContext;
import com.opensymphony.xwork2.ActionInvocation;
import com.opensymphony.xwork2.interceptor.AbstractInterceptor;
public class LoginValidateInterceptor extends AbstractInterceptor{
    @Override
    public String intercept(ActionInvocation invocation) throws Exception {
        // 通过核心调度器invocation来获得调度的Action上下文
        ActionContext actionContext = invocation.getInvocationContext();
        //获得session对象
        Map<String,Object> session = actionContext.getSession();
        //从session中取出名为user的session属性
        String user = (String)session.get("user");
        if(user != null ){
            return invocation.invoke();
        }else{
            //直接返回longin的逻辑视图
            return Action.LOGIN;
        }
    }
}
```

本应用中登录页面login.jsp的代码如下：

```jsp
<body>
    <form action="login.action" method="post">
```

用户名：< input type = "text" name = "uname"/>< br >
密　码：< input type = "password" name = "upass"/>< br >
< input type = "submit" value = "提交">
</form>
</body>

本应用中处理登录的 Action 类的代码如下：

```
package action;
import java.util.Map;
import org.apache.struts2.interceptor.SessionAware;
import com.opensymphony.xwork2.ActionSupport;
public class LoginAction extends ActionSupport implements SessionAware{
    private String uname;
    private String upass;
    private Map<String,Object> session;
    public String getUname() {
        return uname;
    }
    public void setUname(String uname) {
        this.uname = uname;
    }
    public String getUpass() {
        return upass;
    }
    public void setUpass(String upass) {
        this.upass = upass;
    }
    public String execute(){
        if("admin".equals(uname) && "admin".equals(upass) ){
            session.put("user", uname);
            return SUCCESS;
        }
        return LOGIN;
    }
    @Override
    public void setSession(Map<String, Object> arg0) {
        session = arg0;
    }
}
```

本应用中登录成功页面 main.jsp 的代码如下：

< body >
　　< a href = "add.action">添加</a>< br >
　　< a href = "update.action">修改</a>< br >
　　< a href = "delete.action">删除</a>< br >
　　< a href = "query.action">查看</a>< br >
</body>

本应用中处理 add、update、delete 和 query 业务的 Action 代码如下：

```java
package action;
import com.opensymphony.xwork2.ActionSupport;
public class MainAction extends ActionSupport{
    public String add(){
        return SUCCESS;
    }
    public String update(){
        return SUCCESS;
    }
    public String delete(){
        return SUCCESS;
    }
    public String query(){
        return SUCCESS;
    }
}
```

本应用中配置文件 struts.xml 的代码如下：

```xml
<?xml version="1.0" encoding="UTF-8" ?>
<!DOCTYPE struts PUBLIC "-//Apache Software Foundation//DTD Struts Configuration 2.1//EN"
    "http://struts.apache.org/dtds/struts-2.1.dtd">
<struts>
    <package name="ch4" namespace="/" extends="struts-default">
        <interceptors>
            <!-- 定义拦截器 loginInterceptor -->
            <interceptor name="loginInterceptor"
                class="interceptor.LoginValidateInterceptor"></interceptor>
            <!-- 定义拦截器栈 myStack -->
            <interceptor-stack name="myStack">
                <!-- 使用拦截器 loginInterceptor -->
                <interceptor-ref name="loginInterceptor"></interceptor-ref>
                <!-- 使用内置拦截器 -->
                <interceptor-ref name="defaultStack"></interceptor-ref>
            </interceptor-stack>
        </interceptors>
        <global-results>
            <!--考虑到多个 Action 可能都需要跳到 login.jsp,所以定义全局 Result -->
            <result name="login">/login.jsp</result>
        </global-results>
        <!-- 配置登录的 Action -->
        <action name="login" class="action.LoginAction">
            <result>/main.jsp</result>
        </action>
        <!-- 配置 add、update、delete、query 的 Action -->
        <action name="*" class="action.MainAction" method="{1}">
            <interceptor-ref name="myStack"></interceptor-ref>
            <result>/{1}.jsp</result>
        </action>
    </package>
</struts>
```

本应用中 add.jsp、update.jsp、delete.jsp 和 query.jsp 页面的代码略。

## 4.3 本章小结

本章介绍了 Struts 2 的拦截器体系，包括如何配置拦截器、如何定义拦截器，最后给出了一个基于拦截器的权限控制示例，大致演示了拦截器在实际开发中的使用。

## 习 题 4

1. 自定义拦截器类的方式有（   ）。
   A. 实现 Interceptor 接口  　　　　　　B. 实现 AbstractInterceptor 接口
   C. 继承 Interceptor 类  　　　　　　　D. 继承 AbstractInterceptor 类
2. 在 struts.xml 文件中，使用（   ）元素定义拦截器。
   A. <interceptor-ref>  　　　　　　　　B. <interceptor>
   C. <intercept>  　　　　　　　　　　　D. <default-interceptor-ref>
3. Struts 2 主要核心功能是由（   ）实现。
   A. 过滤器  　　　B. 拦截器  　　　C. 类型转换器  　　　D. 配置文件
4. Struts 2 以（   ）为核心，采用（   ）的机制处理用户请求。
   A. Struts  　　　B. WebWork  　　C. 拦截器  　　　　　D. jar 包
5. Struts 2 自定义拦截器中的"return invocation.invoke()"代码表示（   ）。
   A. 不执行目标 Action 的方法，直接返回
   B. 将控制权交给下一个拦截器，或者交给 Action 类的方法
   C. 在自定义拦截器中，该代码是必需的
   D. 以上说法都不对
6. 如下代码，对 test 起作用的拦截器有（   ）。

```
<package name = "default" extends = "struts - default">
    <default - interceptor - ref name = "testInterceptor"/>
    <action name = "test" class = "action.TestAction">
        <interceptor - ref name = "demoInterceptor" />
    </action>
</package>
```

   A. 只有 testInterceptor  　　　　　　　B. 只有 demoInterceptor.
   C. 都不起作用  　　　　　　　　　　　D. 同时起作用
7. 简述 Struts 2 中拦截器的实现原理。
8. 实现自定义拦截器的方法有哪几种？
9. 拦截器与过滤器的区别是什么？

# 表达式语言 OGNL

主要内容

（1）OGNL 基础。
（2）OGNL 基本语法。

OGNL 是 Object Graph Navigation Language（对象图导航语言）的缩写，OGNL 是一种功能强大的表达式语言（Expression Language，EL），通过它简单一致的表达式语法，可以存取对象的任意属性、调用对象的方法、遍历整个对象的结构图、实现字段类型转化等功能。

## 5.1 OGNL 基础

OGNL 称为对象图导航语言。所谓对象图，即以任意一个对象为根，通过 OGNL 可以访问与这个对象关联的其他对象。下面通过一个简单示例讲解对象图。

User.java 的代码如下：

```
public class User {
private String username;
    private Group group;
    public String getUsername() {
        return username;
    }
    public void setUsername(String username) {
        this.username = username;
    }
    public Group getGroup() {
        return group;
    }
    public void setGroup(Group group) {
        this.group = group;
    }
}
```

Group.java 的代码如下：

```
public class Group {
```

```
    private String name;
    private Organization org;
    public String getName() {
        return name;
    }
    public void setName(String name) {
        this.name = name;
    }
    public Organization getOrg() {
        return org;
    }
    public void setOrg(Organization org) {
        this.org = org;
    }
}
```

Organization.java 的代码如下：

```
public class Organization {
    private String orgId;
    public String getOrgId() {
        return orgId;
    }
    public void setOrgId(String orgId) {
        this.orgId = orgId;
    }
}
```

上述 3 个类，描述了通过一个 User 对象，可以导航到 Group 对象，进而导航到 Organization 对象，以 User 对象为根，一个对象图如图 5.1 所示。

在真实的情况下，这个对象图可能会极其复杂，但是通过基本的 getters 方法，都应该能够访问到某个对象的其他关联对象。

下述代码将创建一个 User 对象，及其相关的一系列对象：

```
User(root)
--username
--group
  --name
  --org
    --orgId
```

图 5.1　对象图示例

```
User user = new User();
Group g = new Group();
Organization o = new Organization();
o.setOrgId("ORGID");
g.setOrg(o);
user.setGroup(g);
```

如果通过 Java 代码来进行导航（依赖于 getters 方法），导航到 Organization 的 orgId 属性，如下所示：

```
user.getGroup().getOrg().getOrgId();
```

导航的目的是为了获取某个对象的值、设置某个对象的值或调用某个对象的方法。OGNL 表达式语言的真正目的是在那些不能写 Java 代码的地方执行 Java 代码，或者是更方便地执行 Java 代码。

如果使用 OGNL 表达式来进行导航，导航到 Organization 的 orgId 属性，如下所示：

user.group.org.orgid

可见 OGNL 表达式最大的优点就是："简单"和"直观"，这也是为什么学习 OGNL 表达式的原因。

### 5.1.1 核心知识

**1. OGNL 上下文**

OGNL 表达式的计算都是围绕 OGNL 上下文进行的，OGNL 上下文实际上就是一个 Map 对象，由 ognl.OgnlContext 类（实现了 java.util.Map 接口）表示。

Struts 2 将 OGNL Context 设置为 ActionContext，即在 Struts 2 中 OGNL 上下文 (Context) 的实现为 ActionContext。Struts 2 将 ValueStack（值栈）作为 OGNL 的根对象。

当 Struts 2 接受一个请求时，会迅速创建 ActionContext、ValueStack、Action，然后把 Action 存放进 ValueStack，所以 Action 的实例变量可以被 OGNL 访问。Struts 2 中，OGNL 表达式需要配合 Struts 标签（在后续章节讲解）才可以使用，如：<s:property value="uname"/>。

下面通过一个示例讲解在 Struts 2 中 OGNL 上下文的实现。

1) 编写 Action 类 OGNLContextAction

OGNLContextAction.java 的代码如下：

```java
package action;
import java.util.Date;
import com.opensymphony.xwork2.ActionContext;
import com.opensymphony.xwork2.ActionSupport;
import entity.User;
public class OGNLContextAction extends ActionSupport {
    //Action 的实例变量，必须有 get、set 方法才能使用 OGNL 访问
    private String loginname;
    public String execute() throws Exception{
        loginname = "OGNL 上下文";
        //Struts 2 中 OGNL 上下文(Context)的实现,Context 为一个 Map 对象
        ActionContext context = ActionContext.getContext();
        //往上下文中存放普通属性 uname
        context.put("uname", "cheney");
        //往上下文中存放普通属性对象属性 user
        context.put("user", new User(123, "xxk", 88.9, true, 'B', new Date()));
        return SUCCESS;
    }
    public String getLoginname() {
        return loginname;
    }
    public void setLoginname(String loginname) {
        this.loginname = loginname;
    }
}
```

2) 编写实体类 User

User.java 的代码如下：

```java
package entity;
import java.util.Date;
public class User {
    private int id;
    private String loginname;
    private double score;
    private boolean gender;
    private char cha;
    private Date birthday;
    public User() {
    }
    public User(int id, String loginname, double score, boolean gender,
            char cha, Date birthday) {
        super();
        this.id = id;
        this.loginname = loginname;
        this.score = score;
        this.gender = gender;
        this.cha = cha;
        this.birthday = birthday;
    }
    public int getId() {
        return id;
    }
    public void setId(int id) {
        this.id = id;
    }
    public String getLoginname() {
        return loginname;
    }
    public void setLoginname(String loginname) {
        this.loginname = loginname;
    }
    public double getScore() {
        return score;
    }
    public void setScore(double score) {
        this.score = score;
    }
    public boolean isGender() {
        return gender;
    }
    public void setGender(boolean gender) {
        this.gender = gender;
    }
    public char getCha() {
        return cha;
```

```java
    }
    public void setCha(char cha) {
        this.cha = cha;
    }
    public Date getBirthday() {
        return birthday;
    }
    public void setBirthday(Date birthday) {
        this.birthday = birthday;
    }
    public String info() {
        return "User [birthday=" + birthday + ", cha=" + cha + ", gender="
                + gender + ", id=" + id + ", loginname=" + loginname
                + ", score=" + score + "]";
    }
}
```

3) 配置 Action

在 struts.xml 文件中配置 OGNLContextAction,具体代码如下:

```xml
<action name="ognlContext" class="action.OGNLContextAction">
    <result>/showContext.jsp</result>
</action>
```

4) 编写 JSP 页面 showContext.jsp,显示 OGNL 上下文中的数据

showContext.jsp 页面的代码如下:

```jsp
<%@ page language="java" import="java.util.*" pageEncoding="utf-8"%>
<!-- 使用 Struts 标签 -->
<%@ taglib prefix="s" uri="/struts-tags" %>
<%
    String path = request.getContextPath();
    String basePath = request.getScheme() + "://" + request.getServerName() + ":" + request.getServerPort() + path + "/";
%>
<!DOCTYPE HTML PUBLIC "-//W3C//DTD HTML 4.01 Transitional//EN">
<html>
  <head>
    <base href="<%=basePath%>">
  </head>
  <body>
      <!-- Struts 2 中,OGNL 表达式需要配合 Struts 标签才可以使用 -->
      <!-- Action 中的属性都在值栈中,#可以省略 -->
      <span style="color:#6600cc;">
      访问 Action 中的普通属性
      </span>:<s:property value="loginname"/><br/>
      访问 ActionContext 中的普通属性:
      <s:property value="#uname"/><br/>
      访问 ActionContext 中的对象属性:
      <s:property value="#user.loginname"/><br/>
  </body>
</html>
```

5) 运行 Action

在浏览器地址栏中输入：http://localhost:8080/ch5/ognlContext.action，运行效果如图 5.2 所示。

**2. 值栈**

在 Struts 2 中将 OGNL 上下文设置为 Struts 2 中的 ActionContext，并将值栈作为 OGNL 的根对象。值栈类似于正常的栈，符合后进先出的栈的特点，可以在值栈中放入、删除和查询对象，值栈是 Struts 2 的核心。

图 5.2 显示 OGNL 上下文中的数据

OGNL 设定的根对象（root 对象）在 Struts 2 中就是 ValueStack（值栈）。如果要访问根对象（即 ValueStack）中对象的属性，则可以省略 # 命名空间，直接访问该对象的属性即可。

下面通过一个简单示例讲解值栈。

1) 编写 Action 类，其中 User 类为本节中编写的实体类

ValueStackAction.java 的代码如下：

```java
package action;
import java.util.Date;
import com.opensymphony.xwork2.ActionContext;
import com.opensymphony.xwork2.ActionSupport;
import com.opensymphony.xwork2.util.ValueStack;
import entity.User;
public class ValueStackAction extends ActionSupport{
    public String execute(){
        //获得一个值栈对象
        ValueStack vs = ActionContext.getContext().getValueStack();
        //往值栈压入张三的信息
        User u1 = new User(1, "张三", 88, true, 'B', new Date());
        vs.push(u1);
        //往值栈压入李四的信息
        User u2 = new User(2, "李四", 99, false, 'A', new Date());
        vs.push(u2);
        //往值栈压入王五的信息
        User u3 = new User(3, "王五", 77, false, 'C', new Date());
        vs.push(u3);
        return SUCCESS;
    }
}
```

2) 配置 Action

在 struts.xml 文件中配置 ValueStackAction，具体代码如下：

```xml
<action name="valueStack" class="action.ValueStackAction">
    <result>/showValueStack.jsp</result>
</action>
```

3）编写 JSP 页面 showValueStack.jsp，显示值栈中的数据

showValueStack.jsp 的代码如下：

```jsp
<%@ page language="java" import="java.util.*" pageEncoding="utf-8"%>
<!-- 使用 Struts 标签 -->
<%@taglib prefix="s" uri="/struts-tags" %>
<%
    String path = request.getContextPath();
    String basePath = request.getScheme()+"://"+request.getServerName()+":"+request.getServerPort()+path+"/";
%>
<!DOCTYPE HTML PUBLIC "-//W3C//DTD HTML 4.01 Transitional//EN">
<html>
  <head>
    <base href="<%=basePath%>">
  </head>
  <body>
    <!-- 按照后进先出的方式，取出栈中元素 -->
    <s:property value="#root[0].loginname"/><br>
    <s:property value="#root[1].loginname"/><br>
    <s:property value="#root[2].loginname"/><br>
  </body>
</html>
```

4）运行 Action

在浏览器地址栏中输入：http://localhost:8080/ch5/valueStack.action，运行效果如图 5.3 所示。

**3. OGNL 的访问**

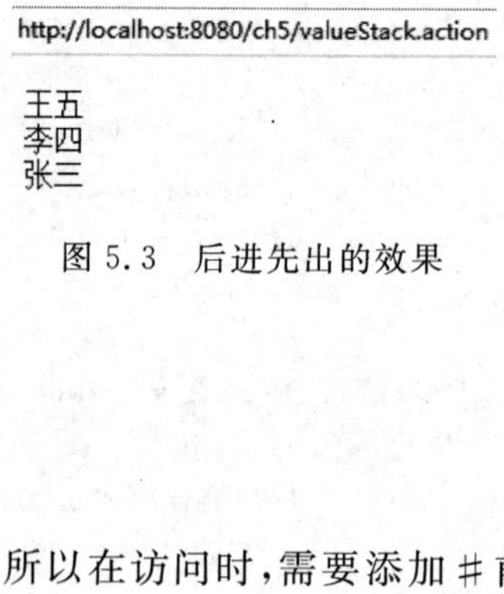

由于 ValueStack 是 Struts 2 中 OGNL 的根对象，如果用户需要访问值栈中的对象，在 JSP 页面可以直接通过下面的 EL 表达式访问 ValueStack 中对象的属性：

`${foo} //获得值栈中某个对象的 foo 属性`

图 5.3　后进先出的效果

或者通过 Struts 标签访问 ValueStack 中对象的属性：

`<s:property value="foo"/>`

如果访问其他 Context 中的对象，由于它们不是根对象，所以在访问时，需要添加#前缀。具体如下。

① application 对象：用于访问 ServletContext，例如#application.userName 或者#application['userName']，相当于调用 ServletContext 的 getAttribute("username")。

② session 对象：用来访问 HttpSession，例如#session.userName 或者#session['userName']，相当于调用 session.getAttribute("userName")。

③ request 对象：用来访问 HttpServletRequest 属性（attribute）的 Map，例如#request.userName 或者#request['userName']，相当于调用 request.getAttribute("userName")。

④ parameters 对象：用于访问 HTTP 的请求参数，例如 #parameters.userName 或者 #parameters['userName']，相当于调用 request.getParameter("username")。

下面是有关 OGNL 访问的示例。

1）编写 Action 类

在 Action 类 OGNLAccessAction 中，使用 request 对象存储属性值，OGNLAccessAction.java 的代码如下：

```java
package action;
import java.util.Map;
import org.apache.struts2.interceptor.RequestAware;
import com.opensymphony.xwork2.ActionSupport;
public class OGNLAccessAction extends ActionSupport implements RequestAware{
    private Map<String, Object> request;
    public String execute(){
        //使用 request 对象存储属性值
        request.put("uname", "OGNL 的访问");
        return SUCCESS;
    }
    @Override
    public void setRequest(Map<String, Object> arg0) {
        request = arg0;
    }
}
```

2）配置 Action

在 struts.xml 文件中配置 OGNLAccessAction，具体代码如下：

```xml
<action name="ognlAccess" class="action.OGNLAccessAction">
    <result>/showOgnlAccess.jsp</result>
</action>
```

3）编写 JSP 页面 showOgnlAccess.jsp，显示 request 对象中的数据

showOgnlAccess.jsp 的代码如下：

```jsp
<%@ page language="java" import="java.util.*" pageEncoding="utf-8"%>
<%
    String path = request.getContextPath();
    String basePath = request.getScheme() + "://" + request.getServerName() + ":" + request.getServerPort() + path + "/";
%>
<!-- 使用 Struts 标签 -->
<%@taglib prefix="s" uri="/struts-tags" %>
<!DOCTYPE HTML PUBLIC "-//W3C//DTD HTML 4.01 Transitional//EN">
<html>
  <head>
    <base href="<%=basePath%>">
  </head>
  <body>
    <!-- 使用 Struts 标签访问 -->
        使用 Struts 标签访问：<br>
```

```
    <s:property value="#request.uname"/><br>
      使用 EL 表达式访问：<br>
<!-- 使用 EL 表达式访问 -->
    ${requestScope.uname}
  </body>
</html>
```

4）运行 Action

在浏览器地址栏中输入：http://localhost:8080/ch5/ognlAccess.action，运行效果如图 5.4 所示。

### 5.1.2 能力目标

理解对象导航图的概念，掌握值栈与 OGNL 的定义与用法。

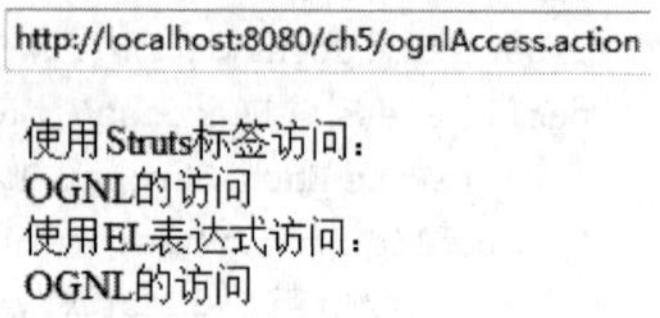

### 5.1.3 任务驱动

图 5.4　OGNL 的访问

1）任务的主要内容

首先，编写一个 Action 类 TaskAction。该 Action 类有一个实例变量 taskString，变量值为"测试任务"；该 Action 类，实现了 SessionAware 接口，并往 session 对象中存放一个字符串值。

其次，编写一个 JSP 页面 task_5_1.jsp，在该页面显示 TaskAction 的实例变量 taskString 的值与 session 对象中的值。

最后，通过"http://localhost:8080/ch5/task.action"演示该任务。

2）任务的代码模板

TaskAction.java 的代码模板如下：

```
package action;
import java.util.Map;
import org.apache.struts2.interceptor.SessionAware;
import com.opensymphony.xwork2.ActionSupport;
public class TaskAction extends ActionSupport implements 【代码1】{
    private static final long serialVersionUID = 1L;
    private String taskString = "测试任务";
    private Map<String, Object> session;
    public String execute(){
        【代码2】//将"我是session中的值"以"myString"存在session对象中
        return SUCCESS;
    }
    @Override
    public void setSession(Map<String, Object> arg0) {
        session = arg0;
    }
    public String getTaskString() {
        return taskString;
    }
    public void setTaskString(String taskString) {
```

```
            this.taskString = taskString;
    }
}
```

Action 的配置代码如下：

```
<action name = "task" class = "action.TaskAction">
    <result>/task_5_1.jsp</result>
</action>
```

task_5_1.jsp 的代码模板如下：

```
<%@ page language = "java" import = "java.util.*" pageEncoding = "UTF-8" %>
<%@ taglib prefix = "s" uri = "/struts-tags" %>
<%
    String path = request.getContextPath();
    String basePath = request.getScheme() + "://" + request.getServerName() + ":" + request.
    getServerPort() + path + "/";
%>
<!DOCTYPE HTML PUBLIC "-//W3C//DTD HTML 4.01 Transitional//EN">
<html>
  <head>
    <base href = "<% = basePath %>">
    <title>My JSP 'task_5_1.jsp' starting page</title>
  </head>
  <body>
        取出 Action 类的实例变量值：<br>
        <s:property value = "【代码 3】"/><br>
        取出 session 对象中的值：<br>
        <s:property value = "【代码 4】"/>
  </body>
</html>
```

3) 任务小结或知识扩展

OGNL 是一种对 Java 对象的 getter 和 setter 属性的表示和绑定语言。通常，OGNL 使用简单一致的表达式语法去存取对象的属性。现在的 Struts 2 中使用 OGNL 取代原来的 EL 来做界面数据绑定。所谓界面数据绑定，也就是把界面元素和对象层某个类的某个属性绑定在一起，修改和显示自动同步。

如果将表达式看作是一个带有语义的字符串，那么 OGNL 无疑成为这个语义字符串与 Java 对象之间沟通的桥梁。其实，只要掌握很少的语法知识，就可以使用 OGNL 实现绝大多数的前端展现功能。

4) 任务代码模板的参考答案

【代码 1】SessionAware

【代码 2】session.put("myString", "我是 session 中的值");

【代码 3】taskString

【代码 4】#session.myString

## 5.1.4 实践环节

创建一个 Web 应用 project514practice.java，具体要求如下。

（1）编写一个 Action 类 OGNLAction，具体代码如下：

```
package action;
import java.util.Map;
import model.TestModel;
import org.apache.struts2.interceptor.SessionAware;
import com.opensymphony.xwork2.ActionContext;
import com.opensymphony.xwork2.ActionSupport;
import com.opensymphony.xwork2.util.ValueStack;
public class OGNLAction extends ActionSupport implements SessionAware{
    private String testString;
    private Map<String, Object> session;
    public String execute(){
        testString = "测试字符串";
        //获得一个值栈对象
        ValueStack vs = ActionContext.getContext().getValueStack();
        //往值栈压入数据
        vs.push("值栈中的数据");
        //创建实体对象
        TestModel tm = new TestModel();
        tm.setId("888888");
        tm.setAge(28);
        //将实体对象存入 session
        session.put("tm", tm);
        return SUCCESS;
    }
    @Override
    public void setSession(Map<String, Object> arg0) {
        session = arg0;
    }
}
```

（2）编写实体类 TestModel.java。

（3）编写 JSP 页面 showInformation.jsp，在该 JSP 页面中使用 OGNL 表达式取出 Action 类中属性 testString 的值、值栈对象中的值以及 session 对象中的值。

（4）配置 Action 并运行。

## 5.2 OGNL 基本语法

OGNL 基本语法是十分简单的，当然 OGNL 支持丰富的表达式，一般情况下，不用担心 OGNL 的复杂性。例如有一个 user 对象，该对象有一个 name 属性，那么使用 OGNL 来获得该 name 属性可以使用如下表达式：user.name。

### 5.2.1 核心知识

**1. 常量**

OGNL 支持的所有常量类型如下。

（1）字符串常量：以单引号或双引号括起来的字符串，例如：'testString' 和

"testString"。在 Java 中,不能用单引号来界定字符串常量,而在 OGNL 中是可以的。不过特别要注意的是如果是单个字符的字符串常量,则必须使用双引号来界定,例如:"S"。OGNL 的字符串也支持转义序列,例如:要在 JSP 页面中输出"You said,"Hello World"。"那么可以使用< s:property >标签,如下:

```
< s:property value = "'You said,\"Hello World\".'"/>
```

(2) 字符常量:以单引号括起来的字符,如:'C'。
(3) 数值常量:Java 中的 int、long、float 和 double。
(4) 布尔常量:true 和 false。
(5) null 常量。

在 JSP 页面中输出常量值的示例代码如下:

```
< s:property value = "'You said,\"Hello World\".'"/><br>
< s:property value = "true"/><br>
< s:property value = "null"/><br>
< s:property value = "12.3f"/>
```

**2. 操作符**

OGNL 支持所有的 Java 操作符(+、-、*、/、++、--、==、!=、=等)与 Java 类似,并提供了一些特有的操作符。与 Java 相同的操作符不再介绍,下面看一下 OGNL 特有的操作符。

1) 逗号(,)或序列操作符

OGNL 的逗号操作符是从 C 语言中借鉴而来的。逗号被用于分隔两个或多个独立的表达式,整个表达式的值是最后一个表达式的值,如:

```
team2.person.name,team1.teamname
```

第一个表达式 team2.person.name 和第二个表达式 team1.teamname,整个表达式的值是第二个表达式的值。

2) 花括号({})操作符

花括号操作符用于创建列表(数组)。使用花括号将元素括起来,元素之间使用逗号分隔,例如:

```
{"zhangsan","lishi","wangwu"}[1]
```

此表达式创建了带有 3 个元素的列表,并且访问其中第二个元素,在 JSP 页面中输出第二个元素值的代码如下:

```
< s:property value = "{'zhangsan','lishi','wangwu'}[1]"/>
```

3) in 和 notin 操作符

in 和 not in 用于判断一个值是否属于一个集合中,如:

```
teamname in {'team1','team2'}
```

此表达式判断 teamname 是否在数组{'team1','team2'}中,在返回 true,不在返回 false。

### 3. OGNL 表达式

使用 OGNL 表达式可以访问属性、方法、静态属性和方法、构造方法、数组以及集合等。

（1）访问属性。OGNL 表达式访问属性的示例代码如下：

访问属性：＜s:property value = "uname"/＞
访问对象属性(get 和 set)：＜s:property value = "user.age"/＞
访问对象属性(get 和 set)：＜s:property value = "employee.manager.uname"/＞

（2）访问方法。OGNL 表达式访问方法的示例代码如下：

访问类对象的方法：＜s:property value = "employee.eat()"/＞
访问 action 的方法：＜s:property value = "add()"/＞

（3）访问静态属性和方法。OGNL 支持调用类中的静态方法和静态字段，可以使用如下语法格式：

@class@method(args)
@class@field

其中 class 必须给出完整的类名，例如：@java.lang.String@valueOf(5)。如果省略 class，那么默认使用的类是 java.lang.Math。示例代码如下：

访问静态方法：＜s:property value = "@com.my.ognl.StaticSample@ma()"/＞
访问静态属性：＜s:property value = "@com.my.ognl.StaticSample@STR"/＞
访问 Math 静态方法：＜s:property value = "@@max(4,7)"/＞

需要注意的是，在配置文件中加上如下常量，OGNL 才能访问静态属性和方法：

＜constant name = "struts.ognl.allowStaticMethodAccess" value = "true"/＞

（4）访问构造方法。OGNL 表达式访问构造方法的示例代码如下：

访问构造方法：＜s:property value = "new com.my.ognl.User(44)"/＞

（5）访问集合和数组。如果需要一个集合元素的时候（例如 List 对象或者 Map 对象），可以使用 OGNL 中同集合相关的表达式，可以使用如下代码直接生成一个 List 对象：

{e1,e2,e3,…}

该 OGNL 表达式中，直接生成了一个 List 对象，该 List 对象中包含 3 个元素：e1、e2 和 e3。如果需要更多的元素，可以按照这样的格式定义多个元素，多个元素之间使用逗号隔开。如下代码可以直接生成一个 Map 对象：

#{key1:value1,key2:value2,…}

Map 类型的集合对象，使用 key-value 格式定义，每个 key-value 元素使用冒号标识，多个元素之间使用逗号隔开。对于集合类型，OGNL 表达式可以使用 in 和 not in 两个元素符号。其中，in 表达式用来判断某个元素是否在指定的集合对象中；not in 判断某个元素是否不在指定的集合对象中。

假设有这样一个 Action 类，其中 User 实体类是 5.1 节中的：

```
package action;
import java.util.ArrayList;
import java.util.Date;
import java.util.HashMap;
import java.util.LinkedHashSet;
import java.util.List;
import java.util.Map;
import java.util.Set;
import org.apache.struts2.ServletActionContext;
import com.opensymphony.xwork2.ActionContext;
import com.opensymphony.xwork2.ActionSupport;
import entity.User;
public class OGNLAction extends ActionSupport {
    private Set<String> courseSet;
    private List<String> list;
    private Map<String,String> map;
    private List<User> userList;
    public String execute() throws Exception{
        //集合 set
        this.courseSet = new LinkedHashSet<String>();
        this.courseSet.add("corejava");
        this.courseSet.add("JSP/Servlet");
        this.courseSet.add("S2SH");
        //集合 List
        this.list = new ArrayList<String>(this.courseSet);
        //集合 Map
        this.map = new HashMap<String, String>();
        this.map.put("x", "xxx");
        this.map.put("y", "yyy");
        this.map.put("z", "zzz");
        //集合 List
        this.userList = new ArrayList<User>();
        this.userList.add(new User(1, "zs", 48.9, true, 'D', new Date()));
        this.userList.add(new User(2, "ls", 68.1, true, 'C', new Date()));
        this.userList.add(new User(3, "ww", 78.2, false, 'B', new Date()));
        this.userList.add(new User(4, "zl", 88.3, true, 'A', new Date()));
        return SUCCESS;
    }
    public Set<String> getCourseSet() {
        return courseSet;
    }
    public void setCourseSet(Set<String> courseSet) {
        this.courseSet = courseSet;
    }
    public List<String> getList() {
        return list;
    }
    public void setList(List<String> list) {
        this.list = list;
```

```java
    }
    public Map<String, String> getMap() {
        return map;
    }
    public void setMap(Map<String, String> map) {
        this.map = map;
    }
    public List<User> getUserList() {
        return userList;
    }
    public void setUserList(List<User> userList) {
        this.userList = userList;
    }
}
```

那么,在上述 Action 转发到的 JSP 页面里可以使用 OGNL 表达式,按如下方式获得集合中的数据:

访问 Action 中的 courseSet 属性: \<s:property value="courseSet"/\>\<br/\>
访问 Action 中的 courseSet 集合中的第一个元素: \<s:property value="courseSet.toArray()[0]"/\>\<br/\>
访问 Action 中的 list 属性: \<s:property value="list"/\>\<br/\>
访问 Action 中的 list 集合中的第一个元素: \<s:property value="list[0]"/\>\<br/\>
访问 Action 中的 userList 属性: \<s:property value="userList"/\>\<br/\>
访问 Action 中的 userList 集合中的第一个元素的 loginname 属性:
    \<s:property value="userList[0].loginname"/\>\<br/\>
访问 Action 中的 map 属性的所有键: \<s:property value="map.keys"/\>\<br/\>
访问 Action 中的 map 属性的所有值: \<s:property value="map.values"/\>\<br/\>
访问 Action 中的 map 属性的指定键对应的值: \<s:property value="map['z']"/\>\<br/\>
访问 Action 中的 map 属性的大小: \<s:property value="map.size"/\>\<br/\>

OGNL 能够引用集合的一些特殊属性,这些属性并不是 JavaBean 模式,例如 size。当表达式引用这些属性时,OGNL 会调用相应的方法,这就是伪属性。

OGNL 中特殊的集合伪属性如表 5.1 所示。

表 5.1  OGNL 中特殊的集合伪属性

| 集合类型 | 伪属性 | OGNL 表达式 | JAVA 代码 |
| --- | --- | --- | --- |
| List,Set,Map | size,isEmpty | list.size,set.isEmpty | list.size(),set.isEmpty() |
| List,Set | iterator | list.iterator,set.iterator | list.iterator(),set.iterator() |
| Map | keys,values | map.keys,map.values | map.keySet(),map.values() |
| Iterator | next,hasNext | it.next,it.hasNext | it.next(),it.hasNext() |

1) 投影

OGNL 提供了一种简单的方式在一个集合中对每一个元素调用相同的方法,或者抽取相同的属性,并将结果保存为一个新的集合,称之为投影。

假设 employees 是一个包含了 employee 对象的列表,那么:

#employees.{name}

表示返回所有雇员的名字的列表。

在投影期间,使用♯this 变量来引用迭代中的当前元素。例如:

objects.{♯this instanceof String? ♯this: ♯this.toString()}

2) 选择

OGNL 提供了一种简单的方式来使用表达式从集合中选择某些元素,并将结果保存到新的集合中,称为选择。选择表达式中关于"?、^、$"特殊字符的使用说明如下。

?:选取匹配逻辑的所有元素。

^:选取匹配选择逻辑的第一个元素。

$:选取匹配的最后一个元素。

选择示例代码如下:

♯employees.{?♯this.salary>3000}

表示将返回薪水大于 3000 的所有雇员的列表。

♯employees.{^♯this.salary>3000}

表示将返回第一个薪水大于 3000 的雇员的列表。

♯employees.{$♯this.salary>3000}

表示将返回最后一个薪水大于 3000 的雇员的列表。

3) OGNL 表达式符号

Struts 2 OGNL 中的♯、%和$符号用法说明如下。

(1) ♯符号的用途一般有 3 种。

① 访问非根对象属性,如♯session.msg 表达式,♯相当于 ActionContext.getContext()。

② 用于选择和投影集合,如 persons.{? ♯this.age>30}。

③ 用来构造 Map,如♯{0:'男',1:'女'}。

(2) %符号。%符号的用途是在标志的属性为字符串类型时,计算 OGNL 表达式的值。如下面的代码所示:

```
<s:set name="foobar" value="♯{'foo1':'bar','foo2':'bar2'}" />
<p>不使用%:<s:url value="♯foobar['foo1']" /></p>
<p>使用%:<s:url value="%{♯foobar['foo1']}" /></p>
```

### 5.2.2 能力目标

掌握 OGNL 表达式的基本语法。

### 5.2.3 任务驱动

1) 任务的主要内容

首先,编写一个工具类 MyUtil,该类有一个静态变量 MAX 和一个静态方法 getStringID()。

其次,编写一个 JSP 页面 task_5_2.jsp,在该页面中使用 OGNL 调用 MyUtil 类中的静态方法和静态变量。

页面 task_5_2.jsp 的运行效果如图 5.5 所示。

2) 任务的代码模板

MyUtil.java 的代码如下：

```java
package util;
import java.text.SimpleDateFormat;
import java.util.Date;
public class MyUtil {
    public static int MAX = 1000;
    /**
     * 获得一个以时间字符串为标准的 ID,固定长度是 17 位
     * @return
     */
    public static String getStringID(){
        String id = null;
        Date date = new Date();
        SimpleDateFormat sdf = new SimpleDateFormat("yyyyMMddHHmmssSSS");
        id = sdf.format(date);
        return id;
    }
}
```

访问静态属性：1000
访问静态方法：20170114112251077

图 5.5　task_5_2.jsp 的运行效果

task_5_2.jsp 的代码模板如下：

```jsp
<%@ page language="java" import="java.util.*" pageEncoding="UTF-8"%>
<%@ taglib prefix="s" uri="/struts-tags" %>
<%
    String path = request.getContextPath();
    String basePath = request.getScheme()+"://"+request.getServerName()+":"+request.getServerPort()+path+"/";
%>
<!DOCTYPE HTML PUBLIC "-//W3C//DTD HTML 4.01 Transitional//EN">
<html>
  <head>
    <base href="<%=basePath%>">
    <title>My JSP 'task_5_2.jsp' starting page</title>
  </head>
  <body>
      访问静态属性：<s:property value="【代码1】"/><br>
      访问静态方法：<s:property value="【代码2】"/>
  </body>
</html>
```

3) 任务小结或知识扩展

需要注意的是，在使用 Struts 2 标签时，如果出现"The Struts dispatcher cannot be found"错误，解决办法是：

首先，在 JSP 页面引入<%@ taglib prefix="s" uri="/struts-tags" %>。

其次，将 WEB-INF 下的 web.xml 中的过滤器配置修改为：

```xml
<filter>
    <filter-name>struts2</filter-name>
    <filter-class>org.apache.struts2.dispatcher.ng.filter.StrutsPrepareAndExecuteFilter
    </filter-class>
</filter>
<filter-mapping>
    <filter-name>struts2</filter-name>
    <url-pattern>/*</url-pattern>
</filter-mapping>
```

4）任务代码模板的参考答案

【代码 1】@util.MyUtil@MAX

【代码 2】@util.MyUtil@getStringID()

### 5.2.4 实践环节

该实践环节是本章的一个综合应用 project524practice，其中 User 实体类是 5.1 节中的。请读懂如下应用的 Action 和 JSP 代码。

OGNLAction.java 的代码如下：

```java
package action;
import java.util.ArrayList;
import java.util.Date;
import java.util.HashMap;
import java.util.LinkedHashSet;
import java.util.List;
import java.util.Map;
import java.util.Set;
import com.opensymphony.xwork2.ActionContext;
import com.opensymphony.xwork2.ActionSupport;
import entity.User;
public class OGNLAction extends ActionSupport {
    private String loginname;
    private String pwd;
    private User user;
    private Set<String> courseSet;
    private List<String> list;
    private Map<String,String> map;
    private List<User> userList;
    public String execute() throws Exception{
        this.loginname = "xkkkkkkkkkkkkkkkkkkkkkkkk";
        this.user = new User(123, "wrr", 88.9, true, 'B', new Date());
        this.courseSet = new LinkedHashSet<String>();
        this.courseSet.add("corejava");
        this.courseSet.add("JSP/Servlet");
        this.courseSet.add("S2SH");
        
        this.list = new ArrayList<String>(this.courseSet);
        this.map = new HashMap<String, String>();
        this.map.put("x", "xxx");
```

```java
            this.map.put("y", "yyy");
            this.map.put("z", "zzz");

            ActionContext context = ActionContext.getContext();
            context.put("uname", "cheney");
            context.put("inte", Integer.valueOf(888888));
            context.put("user2", new User(123, "xxk", 88.9, true, 'B', new Date()));

            this.userList = new ArrayList<User>();
            this.userList.add(new User(1, "zs", 48.9, true, 'D', new Date()));
            this.userList.add(new User(2, "ls", 68.1, true, 'C', new Date()));
            this.userList.add(new User(3, "ww", 78.2, false, 'B', new Date()));
            this.userList.add(new User(4, "zl", 88.3, true, 'A', new Date()));
            context.put("reqAtt", "往ActionContext中put的属性");
            //session
            context.getSession().put("sesAtt", "往ActionContext.getSession()中put的属性");
            //application
            context.getApplication().put("appAtt", "往ActionContext.getApplication()中put的属性");
            return SUCCESS;
        }
        public String getAppName(){
            return "这是OGNL的使用示例代码";
        }
        public String getLoginname() {
            return loginname;
        }
        public void setLoginname(String loginname) {
            this.loginname = loginname;
        }
        public String getPwd() {
            return pwd;
        }
        public void setPwd(String pwd) {
            this.pwd = pwd;
        }
        public User getUser() {
            return user;
        }
        public void setUser(User user) {
            this.user = user;
        }
        public Set<String> getCourseSet() {
            return courseSet;
        }
        public void setCourseSet(Set<String> courseSet) {
            this.courseSet = courseSet;
        }
        public List<String> getList() {
            return list;
        }
```

```java
    public void setList(List<String> list) {
        this.list = list;
    }
    public Map<String, String> getMap() {
        return map;
    }
    public void setMap(Map<String, String> map) {
        this.map = map;
    }
    public List<User> getUserList() {
        return userList;
    }
    public void setUserList(List<User> userList) {
        this.userList = userList;
    }
}
```

struts.xml配置文件的代码如下：

```xml
<?xml version="1.0" encoding="UTF-8"?>
<!DOCTYPE struts PUBLIC "-//Apache Software Foundation//DTD Struts Configuration 2.1//EN"
 "http://struts.apache.org/dtds/struts-2.1.dtd">
<struts>
    <constant name="struts.ognl.allowStaticMethodAccess" value="true"/>
    <package name="my" namespace="/" extends="struts-default">
        <action name="ognl" class="action.OGNLAction">
            <result>/index.jsp</result>
        </action>
    </package>
</struts>
```

JSP 页面 index.jsp 的代码如下：

```jsp
<%@ page language="java" import="java.util.*" pageEncoding="utf-8" %>
<%@taglib prefix="s" uri="/struts-tags" %>
<%
    String path = request.getContextPath();
    String basePath = request.getScheme()+"://"+request.getServerName()+":"+request.getServerPort()+path+"/";
%>
<!DOCTYPE HTML PUBLIC "-//W3C//DTD HTML 4.01 Transitional//EN">
<html>
  <head>
    <base href="<%=basePath%>">
  </head>
  <body>
    <!-- 常量的输出 -->
    <!-- 输出字符串常量,注意里面的单引号,如果没有单引号就是 OGNL 表达式 -->
    <s:property value="'aaaa'"/><br/>
    <s:set name="my" value="'aaaaaaaaaa'"/>
    <s:property value="my"/><br/>
    <!-- Action 中的属性都在值栈中, # 可以省略 -->
```

&lt;span style="color:#6600cc;"&gt;访问 Action 中的普通属性&lt;/span&gt;:&lt;s:property value="loginname"/&gt;&lt;br/&gt;
&lt;span style="color:#ff6600;"&gt;访问 Action 中的对象属性&lt;/span&gt;:
&lt;s:property value="user.birthday"/&gt;&lt;br/&gt;
访问 Action 中的 Set 属性:&lt;s:property value="courseSet.toArray()[0]"/&gt;&lt;br/&gt;
访问 Action 中的 List 属性:&lt;s:property value="list[0]"/&gt;&lt;br/&gt;
访问 Action 中的 Map 属性的键:&lt;s:property value="map.keys.toArray()[1]"/&gt;&lt;br/&gt;
访问 Action 中的 Map 属性的值:&lt;s:property value="map.values.toArray()[1]"/&gt;&lt;br/&gt;
访问 Action 中的 Map 属性的指定键对应的值:&lt;s:property value="map['z']"/&gt;&lt;br/&gt;
访问 Action 中的 Map 属性的大小:&lt;s:property value="map.size"/&gt;&lt;br/&gt;
&lt;hr/&gt;
访问 ActionContext 中的普通属性:&lt;s:property value="#inte"/&gt;&lt;br/&gt;
访问 ActionContext 中的对象属性:&lt;s:property value="#user2.loginname"/&gt;&lt;br/&gt;
&lt;hr/&gt;
&lt;strong&gt;&lt;span style="color:#6600cc;"&gt;访问 Action 中的普通方法:&lt;s:property value="getAppName()"/&gt;&lt;br/&gt;&lt;/span&gt;&lt;/strong&gt;
访问 ActionContext 中的某个对象上的普通方法:&lt;s:property value="#user2.info()"/&gt;&lt;br/&gt;
&lt;hr/&gt;
&lt;span style="color:#33cc00;"&gt;访问静态属性:&lt;s:property value="@java.lang.Math@PI"/&gt;&lt;br/&gt;&lt;/span&gt;
&lt;span style="color:#33cc00;"&gt;访问静态方法:&lt;s:property value="@java.lang.Math@max(45,67)"/&gt;&lt;br/&gt;&lt;/span&gt;
&lt;span style="color:#33cc00;"&gt;访问静态方法:&lt;s:property value="@@min(45,67)"/&gt;&lt;br/&gt;&lt;/span&gt;
&lt;hr/&gt;
&lt;span style="color:#cc0000;"&gt;调用 java.util.Date 的构造方法&lt;/span&gt;:&lt;s:date name="new java.util.Date()" format="yyyy-MM-dd HH:mm:ss"/&gt;&lt;br/&gt;
调用 java.util.Date 的构造方法创建对象,再调用它的方法:&lt;s:property value="new java.util.Date().getTime()"/&gt;&lt;br/&gt;
&lt;hr/&gt;
投影查询:获取 userList 中所有 loginname 的列表:&lt;s:property value="userList.{loginname}"/&gt;&lt;br/&gt;
选择查询:获取 userList 中所有 score 大于 60 的 loginname 列表:&lt;s:property value="userList.{?#this.score>60.0}.{loginname}"/&gt;&lt;br/&gt;
选择查询:获取 userList 中所有 score 大于 60 并且 gender 为 true 的 loginname 列表:&lt;s:property value="userList.{?(#this.score>60.0 && #this.gender)}.{loginname}"/&gt;&lt;br/&gt;
选择查询:获取 userList 中所有 score 大于 60 并且 gender 为 true 的第一个元素的 loginname:&lt;s:property value="userList.{^(#this.score>60.0 && #this.gender)}.{loginname}"/&gt;&lt;br/&gt;
选择查询:获取 userList 中所有 score 大于 60 并且 gender 为 true 的最后一个元素的 loginname:&lt;s:property value="userList.{$(#this.score>60.0 && #this.gender)}.{loginname}"/&gt;&lt;br/&gt;
&lt;hr/&gt;
访问通过 ActionContext 中放入 Request 中的属性:&lt;s:property value="#request.reqAtt"/&gt;&lt;br/&gt;
访问通过 ActionContext 中放入 Session 中的属性:&lt;s:property value="#session.sesAtt"/&gt;&lt;br/&gt;
访问通过 ActionContext 中放入 ServletContext 中的属性:&lt;s:property value="#application.

```
appAtt"/><br/>
            <br/><br/><hr/>
            <s:iterator value = "userList" status = "vs">
                <s:if test = "%{#vs.odd}">
                    <span style = "color: red">
                        <s:property value = "#vs.count"/>:
<s:property value = "loginname"/>,<s:date name = "birthday" format = "yyyy-MM-dd HH:mm:ss"/><br/>
                    </span>
                </s:if>
                <s:else>
                    <span style = "color: blue">
                        <s:property value = "#vs.count"/>:
<s:property value = "loginname"/>,<s:date name = "birthday" format = "yyyy-MM-dd HH:mm:ss"/><br/>
                    </span>
                </s:else>
            </s:iterator>
            <hr/>
        </body>
</html>
```

## 5.3 本章小结

OGNL 类似于 EL 表达式语言,只是一种进行数据访问的语言,但 OGNL 表达式的功能更强大,所以读者应当能够理解 OGNL 表达式,熟悉 OGNL 的各种语法。

# 习 题 5

1. 以下关于 OGNL 的说法正确的是( )。
    A. ActionContext 是 OGNL 的上下文环境
    B. StackContext 是 OGNL 的上下文环境
    C. ValueStack 是 OGNL 的根
    D. ActionContext 是 OGNL 的根

2. 假设在 session 中存在名为 uid 的属性,通过 OGNL 访问该属性,正确的代码是( )。
    A. #uid                           B. #session.uid
    C. uid                            D. ${session.uid}

3. 关于"#session.persions.{? #this.age>20}" OGNL 代码所表示的意义说法正确的是( )。
    A. 从 persons 集合中取出第一个年龄>20 的 Person 对象
    B. 从 persons 集合中取出所有年龄>20 的 Person 对象子集
    C. 从 persons 集合中取出最后一个年龄>20 的 Person 对象

D. 该代码不符合 OGNL 的语法

4. 值栈中有一个字符串对象 userpwd,使用 OGNL 表达式获取字符串长度的方法是（ ）。

  A. ＄{userpwd}

  B. ＄{userpwd.length}

  C. <s:property value="#userpwd.length"/>

  D. <s:property value="userpwd.length()"/>

5. 包 entity 下有一个类 Cat,使用 OGNL 表达式访问 Cat 中的类方法 game 的代码是（ ）。

  A. <s:property value="#entity.Cat.game()"/>

  B. <s:property value="#Cat.game()"/>

  C. <s:property value="entity.Cat.game()"/>

  D. <s:property value="@entity.Cat@game()"/>

6. <s:property value="#session.upass"/>中的#代表（ ）。

  A. 从 ValueStack 取值      B. 从 StackContext 取值

  C. 从 session 取值        D. 从 Action 中取值

7. 请举例说明什么是对象图。

# Struts 2 的标签

**主要内容**

- (1) 非 UI 标签。
- (2) UI 标签。

Struts 2 提供了非常丰富的标签库,包括 UI 标签库、非 UI 标签库以及 Ajax 标签库。Struts 2 标签结合 OGNL 表达式能够完成非常强大的功能。因此,掌握好 Struts 2 常用的标签,并且能够在页面中灵活使用,将起到事半功倍的效果。

Struts 2 标签库相对于 Struts 1 进行了巨大的改进,支持 OGNL 表达式,不再依赖任何表现层技术。Struts 2 标签库可以分为以下 3 类。

### 1. 用户界面标签(UI 标签)

UI 标签,主要用来生成 HTML 元素的标签,又分为表单标签和非表单标签。表单标签,主要用于生成 HTML 页面的 form 元素以及普通表单元素的标签。非表单标签,主要用于在页面生成一些非表单的可视化元素。

### 2. 非用户界面标签(非 UI 标签)

非 UI 标签,主要用于数据访问以及逻辑控制,包括数据访问标签和流程控制标签。数据访问标签,主要用于输出值栈(ValueStack)中的值、完成国际化等功能的标签。流程控制标签,主要用于实现分支、循环等流程控制的标签。

### 3. Ajax 标签

用于支持 Ajax 效果。

## 6.1 数 据 标 签

数据标签主要用于访问值栈中的数据,包含显示一个 Action 里的属性,以及生成国际化输出等功能。主要的数据标签如表 6.1 所示。

表 6.1  数据标签

| 标签名 | 描述 |
| --- | --- |
| action | 用于在 JSP 页面直接调用 Action,当需要调用 Action 时,可以指定 Action 的 name 和 namespace;若指定了 executeResult 参数的值为 true,该标签还会将 Action 的处理结果(视图页面)包含到本页面 |
| bean | 用于创建一个 JavaBean 实例。如果指定了 id 属性,则可以将创建的 JavaBean 实例放入 Stack Context 中 |
| date | 用于格式化输出一个日期 |
| debug | 用于在页面上生成一个调试链接,当单击该链接时,可以看到当前 ValueStack 和 StackContext 中的内容 |
| i18n | 用于指定国际化资源文件的 baseName |
| include | 用于在 JSP 页面中包含其他的 JSP 或 Servlet 资源 |
| param | 用于设置一个参数,通常是用做 bean 标签、url 标签的子标签 |
| push | 用于将某个值放入 ValueStack 的栈顶 |
| set | 用于设置一个新变量,并可以将新变量放入指定的范围内 |
| text | 用于输出国际化消息 |
| url | 用于生成一个 URL 地址 |
| property | 用于生成输出某个值,包括输出 ValueStack、Stack Context 和 Action Context 中的值 |

下面依次介绍这些数据标签,其中 i18n 和 text 标签在国际化章节介绍。

## 6.1.1  核心知识

**1. action 标签**

当需要在 JSP 页面中直接调用 Action 时,可以使用<s:action>标签。该标签有如下几个常用属性。

(1) name:这是一个必填属性,通过该属性指定该标签调用哪个 Action。

(2) namespace:这是一个可选属性,该属性指定该标签调用的 Action 所在的 namespace。

(3) executeResult:这是一个可选属性,该属性指定是否要将 Action 的处理结果页面包含到本页面。该属性值默认是 false,即不包含。

(4) ignoreContextParams:这是一个可选属性,它指定该页面中的请求参数是否需要传入调用的 Action。该参数的默认值是 false,即将本页的请求参数传入被调用的 Action。

下面通过一个示例讲解 action 标签的使用方法。

首先,创建一个名为 ActionTag.java 的 Action 类,代码如下:

```
package action;
import com.opensymphony.xwork2.ActionSupport;
public class ActionTag extends ActionSupport{
    private String paramString = "";
    public String execute(){
        if("".equals(paramString)){
            paramString = "请求参数的值为空!";
        }else{
```

```
            paramString = "请求参数的值为: " + paramString;
        }
        return SUCCESS;
    }
    public String getParamString() {
        return paramString;
    }
    public void setParamString(String paramString) {
        this.paramString = paramString;
    }
}
```

其次,创建两个 JSP 页面,一个是 ActionTag.java 执行后返回的结果页面 action-include.jsp;另一个是调用 Action 的访问页面 actionTag.jsp。

action-include.jsp 的代码如下:

```
<%@ page language="java" import="java.util.*" pageEncoding="utf-8"%>
<%@taglib prefix="s" uri="/struts-tags" %>
<%
    String path = request.getContextPath();
    String basePath = request.getScheme()+"://"+request.getServerName()+":"+request.getServerPort()+path+"/";
%>
<!DOCTYPE HTML PUBLIC "-//W3C//DTD HTML 4.01 Transitional//EN">
<html>
  <head>
    <base href="<%=basePath%>">
  </head>
  <body>
    <!-- 输出 Action 中的 paramString 属性 -->
    <s:property value="paramString"/>
  </body>
</html>
```

actionTag.jsp 的代码如下:

```
<%@ page language="java" import="java.util.*" pageEncoding="utf-8"%>
<%@taglib prefix="s" uri="/struts-tags" %>
<%
    String path = request.getContextPath();
    String basePath = request.getScheme()+"://"+request.getServerName()+":"+request.getServerPort()+path+"/";
%>
<!DOCTYPE HTML PUBLIC "-//W3C//DTD HTML 4.01 Transitional//EN">
<html>
  <head>
    <base href="<%=basePath%>">
  </head>
  <body>
    <h3>s:action 示例</h3>
    将结果包含到本页面中:
```

```
    <s:action name = "actionTag" executeResult = "true" namespace = "/"/>
    <hr>
将结果包含到本页面中,同时阻止本页面的请求参数传入 Action:
    <s:action name = "actionTag" executeResult = "true" namespace = "/" ignoreContextParams = "true"/>
    <hr>
不将结果包含到本页面中:
    <s:action name = "actionTag" executeResult = "false" namespace = "/"/>
  </body>
</html>
```

再次,在 struts.xml 中配置 Action,具体代码如下:

```
<?xml version = "1.0" encoding = "UTF-8" ?>
<!DOCTYPE struts PUBLIC "-//Apache Software Foundation//DTD Struts Configuration 2.1//EN"
    "http://struts.apache.org/dtds/struts-2.1.dtd">
<struts>
    <package name = "ch6" namespace = "/" extends = "struts-default">
        <action name = "actionTag" class = "action.ActionTag">
            <result>/action-include.jsp</result>
        </action>
    </package>
</struts>
```

最后,项目发布后,通过 URL 访问 actionTag.jsp 页面:

http://localhost:8080/ch6/actionTag.jsp?paramString=好好

页面运行效果如图 6.1 所示。

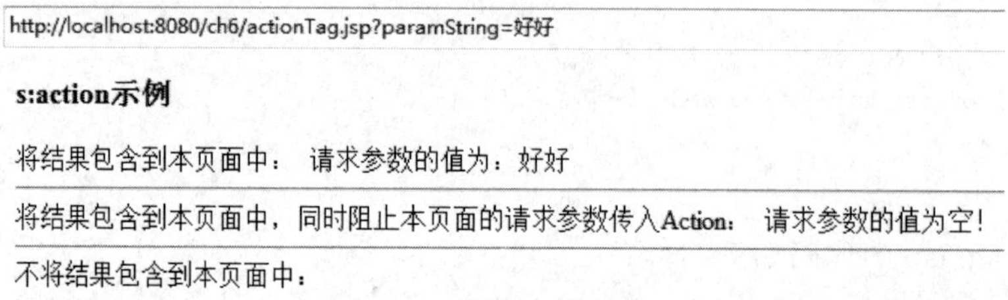

> http://localhost:8080/ch6/actionTag.jsp?paramString=好好
>
> **s:action示例**
>
> 将结果包含到本页面中: 请求参数的值为:好好
> 将结果包含到本页面中,同时阻止本页面的请求参数传入Action: 请求参数的值为空!
> 不将结果包含到本页面中:

图 6.1 使用 action 标签

### 2. bean 标签

bean 标签用于创建一个 JavaBean 的实例。创建 JavaBean 实例时,可以在该标签体内使用 param 标签为 JavaBean 的属性传值,前提是应该为属性提供 set 和 get 方法。使用 bean 标签时可以指定如下两个属性。

(1) name:该属性是一个必填属性,该属性指定要实例化的 JavaBean 的实现类。

(2) var:该属性是一个可选属性。如果指定了该属性,则该 JavaBean 实例会被放入 Stack Context 中(并不是值栈),从而允许直接通过该 var 属性来访问 JavaBean 实例。

**注意**:在 bean 标签体内,bean 标签创建的 JavaBean 实例位于值栈的栈顶,但一旦该 bean 标签结束了,则 bean 标签创建的 JavaBean 实例被移出值栈,除非指定了 var 属性,否则将无法访问该 JavaBean 实例。

下面通过一个简单示例讲解 bean 标签的使用方法。

首先，创建一个 User 类，作为需要实例化的 JavaBean，具体代码如下：

```java
package bean;
public class User {
    private String uname;
    private String uemail;
    public String getUname() {
        return uname;
    }
    public void setUname(String uname) {
        this.uname = uname;
    }
    public String getUemail() {
        return uemail;
    }
    public void setUemail(String uemail) {
        this.uemail = uemail;
    }
}
```

其次，创建 beanTag.jsp，在该 JSP 页面中使用 bean 标签，具体代码如下：

```jsp
<%@ page language="java" import="java.util.*" pageEncoding="utf-8"%>
<%@taglib prefix="s" uri="/struts-tags" %>
<html>
  <head>
    <title>My JSP 'beanTag.jsp' starting page</title>
  </head>
  <body>
      <h3>bean 标签示例</h3>
      <!-- 使用 bean 标签创建 user 对象 -->
      <s:bean name="bean.User" var="user">
          <!-- 通过 param 标签调用 set 方法给 uname、uemail 赋值 -->
          <s:param name="uname" value="'陈恒'"/>
          <!-- 别忘记了单引号'陈恒' -->
          <s:param name="uemail" value="'123456@qq.com'"/>
          <!-- 使用 property 标签输出 JavaBean 的属性值 -->
          在 bean 标签体内取 JavaBean 的属性值：<br>
          姓名为：<s:property value="uname"/><br>
          邮箱为：<s:property value="uemail"/><br>
      </s:bean>
      <br>
      使用 var 属性在 bean 标签体外取 JavaBean 的属性值：<br>
      姓名为：<s:property value="#user.uname"/><br>
      邮箱为：<s:property value="#user.uemail"/><br>
  </body>
</html>
```

最后,运行 beanTag.jsp 页面,效果如图 6.2 所示。

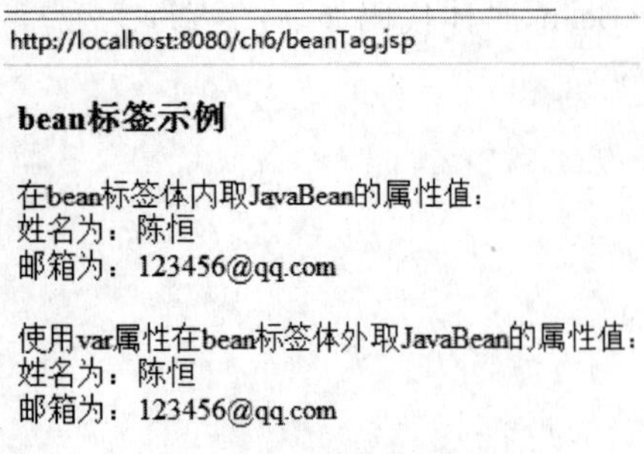

**3. date 标签**

date 标签用于格式化输出一个日期。除了可以直接格式化输出一个日期外,date 标签还可以计算指定日期和当前时刻之间的时差。使用 date 标签时可以指定如下几个属性。

(1) format:这是一个可选属性,如果指定了该属性,将根据该属性指定的格式来格式化日期。

(2) name:这是一个必填属性,该属性指定要格式化的日期值。

图 6.2 使用 bean 标签

(3) nice:这是一个可选属性,该属性只能为 true 或 false,它用于指定是否输出指定日期和当前时刻之间的时差。该属性默认是 false,即表示不输出时差。

(4) var:这是一个可选属性,该属性指定引用该元素的 var 值。

下面通过一个简单示例讲解 date 标签的使用方法。

首先,创建一个 JSP 页面 dateTag.jsp,具体代码如下:

```
<%@ page language="java" import="java.util.*" pageEncoding="utf-8"%>
<%@taglib prefix="s" uri="/struts-tags" %>
<html>
  <head>
    <title>My JSP 'dateTag.jsp' starting page</title>
  </head>
  <body>
    <%
        //距离格林治时间(1,1970,00:00:00 GMT)的毫秒是 1888888888 的日期
        Date date = new Date(1888888888);
        //将 date 实例设置成 pageContext 里的属性
        pageContext.setAttribute("date", date);
    %>
    当前日期,不指定 format:
    <s:date name="new java.util.Date()"/>
    <br>
    当前日期,指定 format:
    <s:date name="new java.util.Date()" format="yyyy.MM.dd"/>
    <hr>
    生成日期:
    <s:date name="#attr.date" nice="false" format="yyyy.MM.dd"/>
    <!-- 如果 nice="true" 和 format 属性都指定了,则会输出指定日期和当前日期的时差,而 format 失效 -->
    <br>
    距今已有:
    <s:date name="#attr.date" nice="true"/>
  </body>
</html>
```

其次，运行 JSP 页面，效果如图 6.3 所示。

### 4. debug 标签

debug 标签主要用于辅助测试，在页面生成一个链接，通过该链接可以查看到 ValueStack 和 Stack Context 中所有的值信息。

在"2. bean 标签"的 beanTag.jsp 页面中增加<s:debug/>标签，通过浏览器运行该页面将看到如图 6.4 所示的页面。

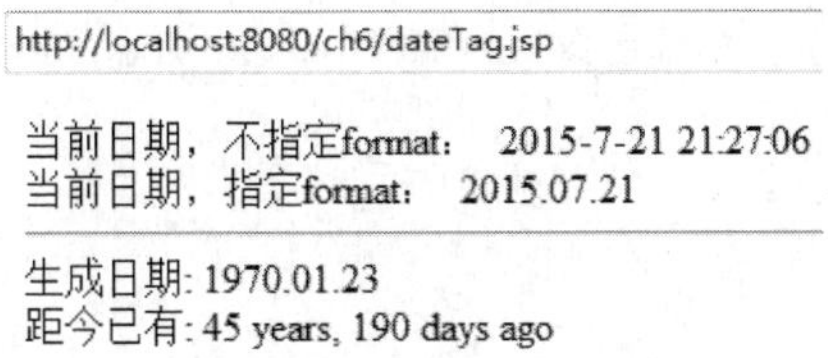

图 6.3　使用 date 标签 　　　　图 6.4　使用 debug 标签

单击图 6.4 中的"[Debug]"链接，将看到如图 6.5 所示的页面。

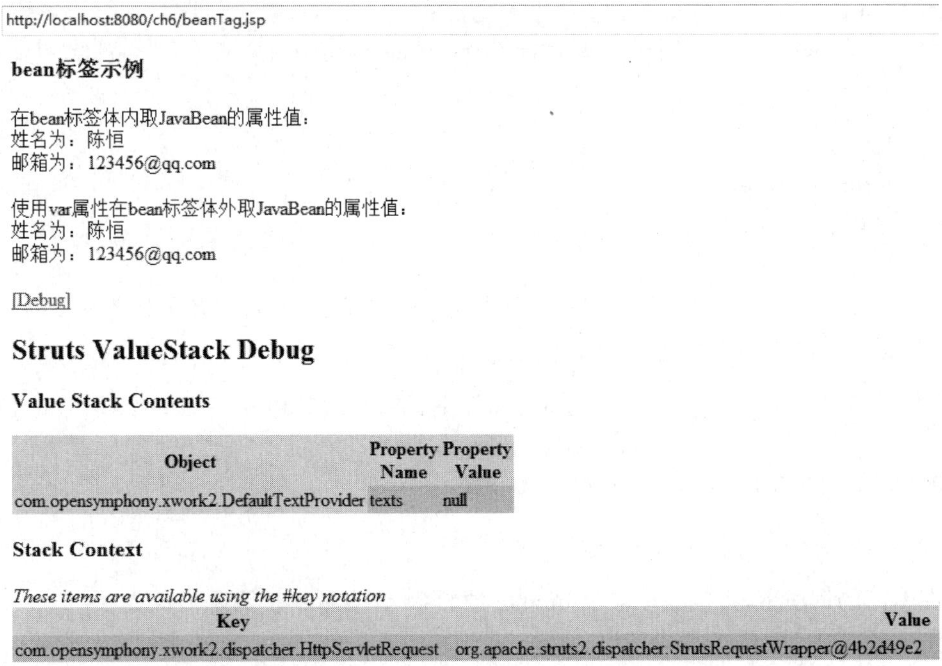

图 6.5　Struts 2 的调试页面

### 5. include 标签

include 标签用于将一个 JSP 页面或 Servlet 包含到本页面。使用该标签时必须提供一个 value 属性，表示需要包含的 JSP 页面或 Servlet。

下面通过一个简单示例讲解 include 标签的使用方法。

首先,创建一个 includeTag.jsp 页面,作为主页面,具体代码如下:

```jsp
<%@ page language="java" import="java.util.*" pageEncoding="utf-8"%>
<%@taglib prefix="s" uri="/struts-tags" %>
<html>
  <head>
    <title>My JSP 'includeTag.jsp' starting page</title>
  </head>
  <body>
    本页面的内容
    <hr>
    <s:include value="include-page.jsp">
     <!-- 使用 param 标签传参到 include-page.jsp,如果参数值为字符串常量,别忘记加'' -->
     <s:param name="uname" value="'陈恒'"/>
    </s:include>
  </body>
</html>
```

其次,创建被包含页面 include-page.jsp,具体代码如下:

```jsp
<%@ page language="java" import="java.util.*" pageEncoding="utf-8"%>
<html>
  <head>
    <title>My JSP 'include-page.jsp' starting page</title>
  </head>
  <body>
    参数为:
    ${param.uname}
  </body>
</html>
```

最后,运行 includeTag.jsp 页面,效果如图 6.6 所示。

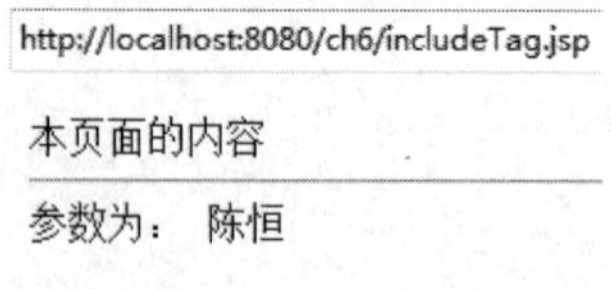

图 6.6 使用 include 标签

### 6. param 标签

param 标签主要用于为其他标签提供参数,它的使用方式主要有两种。

第一种用法:

```
<s:param name="uname">陈恒</s:param>
```

在上面的用法中,指定一个名为 uname 的参数,该参数的值为陈恒。

第二种用法:

```
<s:param name="uname" value="gogo"/>
```

在上面的用法中,指定一个名为 uname 的参数,该参数的值为 gogo 对象的值。如果 gogo 对象不存在,则 uname 参数的值为 null。如果想指定 uname 参数的值为 gogo 字符串,则应该这样写:

```
<s:param name = "uname" value = "'gogo'"/>
```

有关 param 标签的示例，前面已经讲了很多，在这里不再赘述。

### 7. push 标签

push 标签用于将某个值放到值栈的栈顶，从而可以更简单地访问该值。使用该标签时，有个必选属性 value，value 属性是需要放到栈顶的值。

下面 JSP 页面 pushTag.jsp 将一个值放到值栈的栈顶，从而可以通过<s:property>标签访问。具体代码如下：

```
<%@ page language = "java" import = "java.util.*" pageEncoding = "utf-8"%>
<%@taglib prefix = "s" uri = "/struts-tags" %>
<html>
  <head>
    <title>My JSP 'pushTag.jsp' starting page</title>
  </head>
  <body>
    <h3>push 标签示例</h3>
    <!-- 使用 bean 标签创建 user 对象 -->
    <s:bean name = "bean.User" var = "user">
        <!-- 通过 param 标签调用 set 方法给 uname、uemail 赋值 -->
        <s:param name = "uname" value = "'陈恒'"/>
        <!-- 别忘记了单引号'陈恒' -->
        <s:param name = "uemail" value = "'123456@qq.com'"/>
    </s:bean>
    <!-- 用 push 标签将 Stack Context 中的 user 实例放入 ValueStack 栈顶 -->
    <s:push value = "#user">
        <!-- 输出栈顶元素的属性 -->
        <s:property value = "uname"/>
        <s:property value = "uemail"/>
    </s:push>
  </body>
</html>
```

### 8. set 标签

set 标签用于将某个值放入指定范围内。使用 set 标签可以理解为定义一个新变量，且将一个已有的值赋值给新变量，并且可以将新变量放到指定的范围内。使用 set 标签有如下属性。

（1）scope：这是一个可选属性，指定新变量被放置的范围，该属性可选的值为 application、session、request、page 或 action。如果没有指定属性，则默认设置在 Stack Context 中。

（2）value：这是一个可选属性，指定将赋给变量的值。如果没有指定该属性，则将 ValueStack 栈顶的值赋给新变量。

（3）var：这是一个可选属性，重新生成的新变量的名字，从而允许直接通过此 var 来访问。

下面 JSP 页面 setTag.jsp 先定义了一个 JavaBean 实例，然后通过 set 标签将该

JavaBean 实例放入指定范围。具体代码如下：

```jsp
<%@ page language="java" import="java.util.*" pageEncoding="utf-8"%>
<%@taglib prefix="s" uri="/struts-tags" %>
<html>
  <head>
    <title>My JSP 'setTag.jsp' starting page</title>
  </head>
  <body>
      <h3>set 标签示例</h3>
      <!-- 使用 bean 标签创建 user 对象 -->
      <s:bean name="bean.User" var="user">
            <!-- 通过 param 标签调用 set 方法给 uname、uemail 赋值 -->
            <s:param name="uname" value="'陈恒'"/>
            <!-- 别忘记了单引号'陈恒' -->
            <s:param name="uemail" value="'123456@qq.com'"/>
      </s:bean>
      <!-- 将 user 对象放入 Page 范围 -->
      <s:set var="u1" value="#user" scope="page"/>
      <!-- 将 user 对象放入 request 范围 -->
      <s:set var="u2" value="#user" scope="request"/>
      <!-- 将 user 对象放入 session 范围 -->
      <s:set var="u3" value="#user" scope="session"/>
      <!-- 将 user 对象放入 application 范围 -->
      <s:set var="u4" value="#user" scope="application"/>
      <!-- 将 user 对象放入默认范围 -->
      <s:set var="u5" value="#user"/>
      取出 Page 范围的值：
      <s:property value="#attr.u1.uname"/><br>
      <!-- 也可以使用 ${pageScope.u1.uname}取 -->
      取出 request 范围的值：
      <s:property value="#request.u2.uname"/><br>
      <!-- 也可以使用 ${requestScope.u2.uname}取 -->
      取出 session 范围的值：
      <s:property value="#session.u3.uname"/><br>
      取出 application 范围的值：
      <s:property value="#application.u4.uname"/><br>
      取出默认范围(Stack Context)的值：
      <s:property value="#u5.uname"/><br>
  </body>
</html>
```

**9. url 标签**

url 标签用于生成一个 URL 地址，可以在 url 标签中嵌套 param 标签来提供额外的参数。该标签有如下几个常用属性。

（1）value：这是一个可选属性，指定生成 URL 的地址值，如果 value 不提供就用 action 属性指定的 Action 作为 URL 地址。

(2) action：这是一个可选属性，指定生成 URL 的地址为哪个 Action，如果 Action 不提供，就使用 value 作为 URL 的地址值。

(3) namespace：这是一个可选属性，该属性指定命名空间。

(4) method：这是一个可选属性，指定使用 Action 的方法。

需要注意的是，action 属性和 value 属性的作用大致相同，只是 action 指定的是一个 Action，因此系统会自动在 action 指定的属性后添加 .action 的后缀。只要指定 action 和 value 两个属性之一即可，如果两个属性都没有指定，就以当前页面作为 URL 的地址值。

下面通过一个简单示例讲解 url 标签的使用方法。

首先，创建 JSP 页面 urlTag.jsp，具体代码如下：

```
<%@ page language = "java" import = "java.util.*" pageEncoding = "utf-8" %>
<%@ taglib prefix = "s" uri = "/struts-tags" %>
<html>
  <head>
    <title>My JSP 'urlTag.jsp' starting page</title>
  </head>
  <body>
    只指定 value 属性：
    <s:url value = "actionTag.action"/><br>
    指定 action 属性，且使用 param 传入参数的形式：
    <s:url action = "actionTag">
      <s:param name = "uname" value = "'chenheng'"/>
    </s:url>
    <br>
    <s:set var = "myurl" value = "'www.baidu.com'"/>
    只指定 value 属性，属性为字符串：
    <s:url value = "%{#myurl}"/>
  </body>
</html>
```

其次，运行 JSP 页面，效果如图 6.7 所示。

```
http://localhost:8080/ch6/urlTag.jsp

只指定value属性：    actionTag.action
指定action属性，且使用param传入参数的形式：  /ch6/actionTag.action?uname=chenheng
只指定value属性，属性为字符串：   www.baidu.com
```

图 6.7 使用 url 标签

例如，下面一段程序是分页常用的代码，在该代码中使用了 url 和 href 标签：

```
<s:url id = "url_pre" value = "card/queryCard.action">
    <s:param name = "pageCur" value = "pageCur-1"></s:param>
</s:url>
<s:url id = "url_next" value = "card/queryCard.action">
    <s:param name = "pageCur" value = "pageCur+1"></s:param>
</s:url>
```

```
<s:a href="%{url_pre}">上一页</s:a>
<s:a href="%{url_next}">下一页</s:a>
```

#### 10. property 标签

property 标签用于输出指定值。property 标签输出 value 属性指定的值，如果没有指定 value 属性，则默认输出 ValueStack 栈顶的值。前面包含了大量使用 property 标签的示例，此处不再赘述。

### 6.1.2 能力目标

掌握常用的数据标签的使用方法。

### 6.1.3 任务驱动

1) 任务的主要内容

编写一个 JSP 页面 task_6_1.jsp。首先，使用 bean 标签和 6.1.1 节的 User 类实例化 JavaBean 对象 myUser，同时在 bean 标签体内使用 param 标签调用 set 方法给 uname、uemail 赋值；其次，使用 set 标签将对象 myUser 赋值给一个新对象 yourUser，有效范围是 request；最后，使用 property 标签输出对象 yourUser 的属性值。

2) 任务的代码模板

task_6_1.jsp 的代码模板如下：

```jsp
<%@ page language="java" import="java.util.*" pageEncoding="UTF-8"%>
<%@taglib prefix="s" uri="/struts-tags" %>
<%
    String path = request.getContextPath();
    String basePath = request.getScheme()+"://"+request.getServerName()+":"+request.getServerPort()+path+"/";
%>
<!DOCTYPE HTML PUBLIC "-//W3C//DTD HTML 4.01 Transitional//EN">
<html>
  <head>
    <base href="<%=basePath%>">
    <title>My JSP 'task_6_1.jsp' starting page</title>
  </head>
  <body>
    <!-- 代码1 使用bean标签和6.1.1节的User类实例化JavaBean对象myUser -->
    【代码1】
        <!-- 通过param标签调用set方法给uname、uemail赋值 -->
        <s:param name="uname" value="'陈恒'"/>
        <!-- 别忘记了单引号'陈恒' -->
        <s:param name="uemail" value="'123456@qq.com'"/>
    </s:bean>
    <!-- 代码2 使用set标签将对象myUser赋值给一个新对象yourUser,有效范围是request -->
    【代码2】
```

```
        <!-- 代码 3 使用 property 标签输出对象 yourUser 的属性 uname 的值 -->
        【代码 3】<br>
        <!-- 代码 4 使用 property 标签输出对象 yourUser 的属性 uemail 的值 -->
        【代码 4】<br>
    </body>
</html>
```

3) 任务小结或知识扩展

在 JSP 页面中使用 Struts 2 标签时,首先在 JSP 代码的顶部加入以下代码:

```
<%@taglib prefix="s" uri="/struts-tags" %>
```

另外,需要注意的是,在使用 Struts 2 标签时,如果出现"The Struts dispatcher cannot be found"错误,解决办法是:

首先,在 JSP 页面引入<%@ taglib prefix="s"  uri="/struts-tags" %>。

其次,将 WEB-INF 下的 web.xml 中的过滤器配置修改为:

```
<filter>
    <filter-name>struts2</filter-name>
    <filter-class>org.apache.struts2.dispatcher.ng.filter.StrutsPrepareAndExecuteFilter</filter-class>
</filter>
<filter-mapping>
    <filter-name>struts2</filter-name>
    <url-pattern>/*</url-pattern>
</filter-mapping>
```

4) 任务代码模板的参考答案

【代码 1】`<s:bean name="bean.User" var="myUser">`

【代码 2】`<s:set var="yourUser" value="#myUser" scope="request"/>`

【代码 3】`<s:property value="#request.yourUser.uname"/>`

【代码 4】`<s:property value="#request.yourUser.uemail"/>`

### 6.1.4 实践环节

编写 JSP 页面 practice614.jsp,具体要求如下。

(1) 使用 bean 标签和 6.1.1 节的 User 类实例化 JavaBean 对象 user,同时在 bean 标签体内使用 param 标签调用 set 方法给 uname、uemail 赋值。然后,在 bean 标签体外使用 property 标签输出 JavaBean 的属性值。

(2) 使用 date 标签生成格式为"dd-MM-yyyy"的当前日期,并使用 date 标签计算 1990.10.10 距离当今多少年多少天。

(3) 使用 set 标签将一个字符串值分别存入 application、session、request、page 范围内,并使用 property 标签分别输出各范围内的该字符串。

(4) 使用 debug 标签调试该页面。

(5) 页面运行效果如图 6.8 所示。

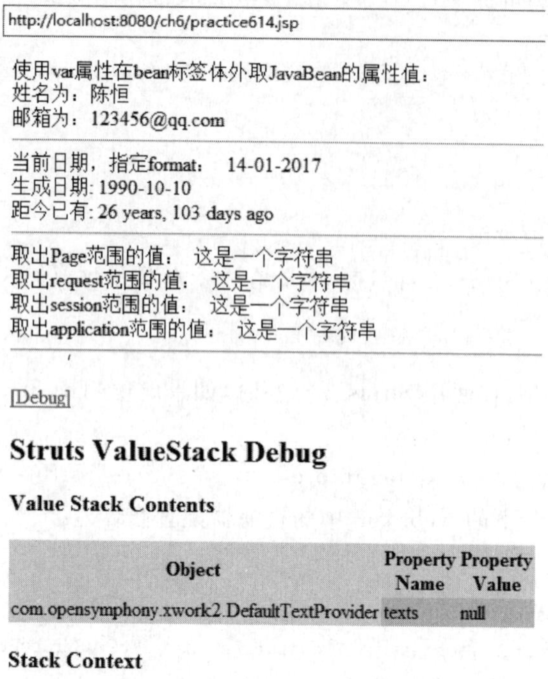

```
http://localhost:8080/ch6/practice614.jsp
```

使用var属性在bean标签体外取JavaBean的属性值：
姓名为：陈恒
邮箱为：123456@qq.com

当前日期，指定format： 14-01-2017
生成日期：1990-10-10
距今已有：26 years, 103 days ago

取出Page范围的值： 这是一个字符串
取出request范围的值： 这是一个字符串
取出session范围的值： 这是一个字符串
取出application范围的值： 这是一个字符串

[Debug]

**Struts ValueStack Debug**

**Value Stack Contents**

| Object | Property Name | Property Value |
|---|---|---|
| com.opensymphony.xwork2.DefaultTextProvider texts | | null |

**Stack Context**

图6.8　practice614.jsp 页面效果

## 6.2　流程控制标签

流程控制标签可以完成输出流程控制，例如分支、循环等操作，也可以完成对集合的合并、排序等操作。主要流程控制标签如表6.2所示。

表6.2　主要流程控制标签

| 标　签　名 | 描　　述 |
|---|---|
| if/elseif/else | 用于控制选择输出的标签 |
| append | 用于将多个集合拼接成一个新的集合 |
| generator | 用于将一个字符串解析成一个集合 |
| iterator | 用于将集合迭代输出 |
| merge | 用于将多个集合拼接成一个新的集合，但与append拼接方式不同 |
| sort | 用于对集合进行排序 |
| subset | 用于截取集合的部分元素，形成新的子集合 |

下面依次介绍这些流程控制标签。

### 6.2.1　核心知识

**1. if/elseif/else 标签**

if/elseif/else 这3个标签可以在页面中使用，类似于Java代码中的 if…else if…else 结构，都是用于分支控制的，它们根据一个 boolean 表达式的值，来决定是否计算、输出标签体的内容。

这 3 个标签可以组合使用,只有 if 标签可以单独使用,后面的 elseif 和 else 都不可单独使用,必须与 if 标签结合使用,其中 if 标签可以与多个 elseif 标签结合使用,并可以结合一个 else 标签使用。

对于 if 标签和 elseif 标签必须指定一个 test 属性,该属性就是进行条件判断的逻辑表达式。下面来看一个用于显示成绩结果的 JSP 页面 ifTag.jsp,具体代码如下:

```jsp
<%@ page language="java" import="java.util.*" pageEncoding="utf-8"%>
<%@taglib prefix="s" uri="/struts-tags" %>
<html>
  <head>
    <title>My JSP 'ifTag.jsp' starting page</title>
  </head>
  <body>
    <h3>if/elseif/else 示例</h3>
    <!-- 使用 set 标签来定义一个变量 score,值为 95 -->
    <s:set var="score" value="95"/>
    考试分数:
    <s:property value="#score"/><br>
    考试等级:
    <!-- 使用 if/elseif/else 标签判断等级 -->
    <s:if test="#score>=90">优秀</s:if>
    <s:elseif test="#score>=80">良好</s:elseif>
    <s:elseif test="#score>=70">中等</s:elseif>
    <s:elseif test="#score>=60">及格</s:elseif>
    <s:else>不及格</s:else>
  </body>
</html>
```

页面 ifTag.jsp 运行效果如图 6.9 所示。

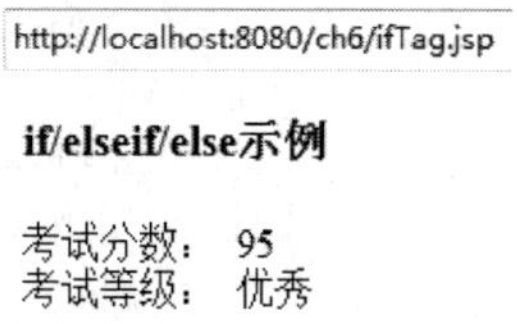

**2. iterator 标签**

iterator 标签主要用于对集合(List、Set、Map 和数组)进行迭代。使用该标签时可以指定如下属性。

(1) value:这是一个可选的属性,该属性指定被迭代的集合,被迭代的集合通常使用 OGNL 表达式指定。如果没有指定该属性,则使用 ValueStack 栈顶的集合。

图 6.9 使用 if/elseif/else 标签

(2) var:这是一个可选的属性,临时存储 value 集合里元素的值。

(3) status:这是一个可选的属性,该属性指定迭代时的 IteratorStatus 实例,通过该实例即可判断当前迭代元素的属性。假如 status 属性值为 st,则可以通过"#st.count"返回当前迭代元素的数量;通过"#st.index"返回当前迭代元素的索引;通过"#st.even"返回当前迭代元素的索引是否为偶数;通过"#st.odd"返回当前迭代元素的索引是否是奇数;通过"#st.first"返回当前迭代元素是否是第一个元素;通过"#st.last"返回当前迭代元素是否是最后一个元素。

(4) step:这是一个可选的属性,该属性值指定迭代的步长。

(5) begin:这是一个可选的属性,该属性指定集合迭代的开始下标。

（6）end：这是一个可选的属性，该属性指定集合迭代的结束下标。

下面创建一个 JSP 页面 iteratorTag.jsp，来进行迭代操作，具体代码如下：

```jsp
<%@ page language="java" import="java.util.*" pageEncoding="utf-8"%>
<%@ taglib prefix="s" uri="/struts-tags" %>
<html>
  <head>
    <title>My JSP 'iteratorTag.jsp' starting page</title>
  </head>
  <body>
    <h3>iterator 标签示例</h3>
    迭代 List：<br>
    <table border="1">
        <tr>
            <th>序号</th>
            <th>书名</th>
        <tr>
        <s:iterator value="{'HTML 与 CSS 网页设计教学做一体化教程','JSP 网站设计教学做一体化教程','软件工程教学做一体化教程','基于 Eclipse 平台的 JSP 应用教程','Struts 2 框架应用教程'}"
            var="book" status="st">
            <!-- 迭代输出 -->
            <tr>
                <!-- 使用 status 属性找索引，索引从 0 开始 -->
                <td><s:property value="#st.index + 1"/></td>
                <td><s:property value="book"/></td>
            </tr>
        </s:iterator>
    </table>
    <hr>
    迭代 Map：<br>
    <table border="1">
        <tr>
            <th>书名</th>
            <th>作者</th>
        <tr>
        <s:iterator value="#{'HTML 与 CSS 网页设计教学做一体化教程':'陈恒',
        'JSP 网站设计教学做一体化教程':'陈恒','软件工程教学做一体化教程':'陈恒',
        '基于 Eclipse 平台的 JSP 应用教程':'陈恒','Struts 2 框架应用教程':'陈恒'}">
            <tr>
            <!-- 输出 Map 对象里的 key -->
            <td><s:property value="key"/></td>
            <!-- 输出 Map 对象里的 value -->
            <td><s:property value="value"/></td>
            </tr>
        </s:iterator>
    </table>
  </body>
</html>
```

页面 iteratorTag.jsp 的运行效果如图 6.10 所示。

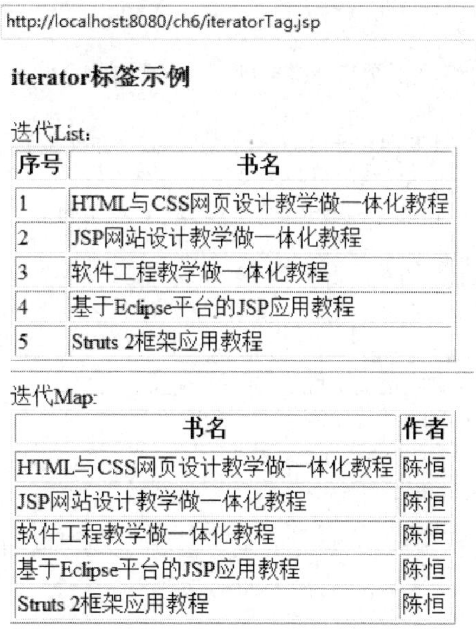

图 6.10　使用 iterator 标签

### 3. append 标签

append 标签用于将多个集合对象拼接起来，组成一个新的集合。使用 append 标签时需要指定一个 var 属性，该属性确定拼接生成新集合的名字。除此之外，append 标签可以接受多个 param 子标签，每个子标签指定一个集合，append 标签负责将 param 标签指定的多个集合拼接成一个集合。

下面创建一个 JSP 页面 appendTag.jsp，来进行拼接操作，具体代码如下：

```
<%@ page language = "java" import = "java.util.*" pageEncoding = "utf-8" %>
<%@ taglib prefix = "s" uri = "/struts-tags" %>
<html>
  <head>
    <title>My JSP 'appendTag.jsp' starting page</title>
  </head>
  <body>
    <h3>append 标签示例</h3>
    拼接并迭代 List:<br>
    <!-- 使用 append 标签将多个集合拼接成一个新集合 -->
    <s:append var = "newList">
      <s:param value = "{'HTML 与 CSS 网页设计教学做一体化教程','JSP 网站设计教学做一体化教程'}"/>
      <s:param value = "{'软件工程教学做一体化教程','基于 Eclipse 平台的 JSP 应用教程','Struts 2 框架应用教程'}"/>
    </s:append>
    <!-- 迭代新集合 newList -->
    <s:iterator var = "book" value = "newList" status = "st">
      <s:property value = "#st.index + 1"/> 
      <s:property value = "book"/><br>
    </s:iterator>
```

```
    <hr>
    拼接并迭代 Map:<br>
    <!-- 使用 append 标签将多个集合拼接成一个新集合 -->
    <s:append var="newMap">
      <s:param value="#{'HTML 与 CSS 网页设计教学做一体化教程':'陈恒',
        'JSP 网站设计教学做一体化教程':'陈恒'}"/>
      <s:param value="#{'软件工程教学做一体化教程':'陈恒',
        '基于 Eclipse 平台的 JSP 应用教程':'陈恒',
        'Struts 2 框架应用教程':'陈恒'}"/>
    </s:append>
    <!-- 迭代新集合 newMap -->
    <s:iterator var="book" value="newMap" status="st">
      <!-- 输出 Map 对象里的 key -->
      <s:property value="key"/> 
      <!-- 输出 Map 对象里的 value -->
      <s:property value="value"/><br>
    </s:iterator>
  </body>
</html>
```

页面 appendTag.jsp 的运行效果如图 6.11 所示。

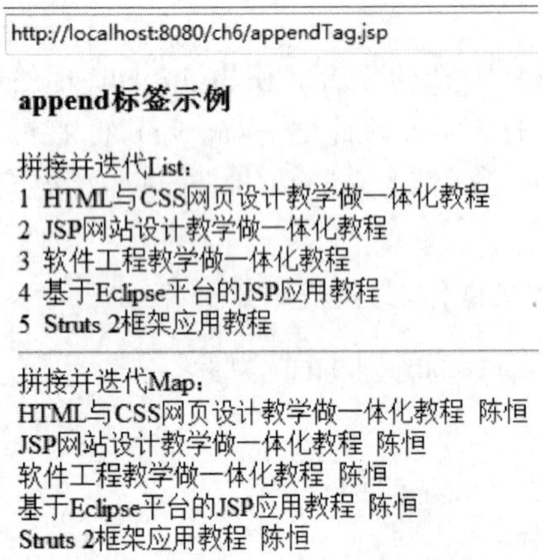

**append标签示例**

拼接并迭代List:
1 HTML与CSS网页设计教学做一体化教程
2 JSP网站设计教学做一体化教程
3 软件工程教学做一体化教程
4 基于Eclipse平台的JSP应用教程
5 Struts 2框架应用教程

拼接并迭代Map:
HTML与CSS网页设计教学做一体化教程 陈恒
JSP网站设计教学做一体化教程 陈恒
软件工程教学做一体化教程 陈恒
基于Eclipse平台的JSP应用教程 陈恒
Struts 2框架应用教程 陈恒

图 6.11  使用 append 标签

### 4. generator 标签

使用 generator 标签可以将指定的字符串分隔成多个字串,临时生成的字串可以使用 iterator 标签来迭代输出。在该标签的标签体内,整个临时生成的集合将位于 ValueStack 的顶端,一旦该标签结束,该集合将被移出 ValueStack。使用该标签时可以指定如下属性。

(1) var:这是一个可选属性,如果指定了该属性,则将生成的字串集合放在 PageContext 属性中。

(2) val:这是一个必填属性,该属性指定被解析的字符串。

(3) separator:这是一个必填属性,该属性指定用于解析字符串的分隔符。

(4) count:这是一个可选属性,该属性指定生成子字符串集合中元素的总数。

下面创建一个 JSP 页面 generatorTag.jsp,来进行解析操作,具体代码如下:

```
<%@ page language="java" import="java.util.*" pageEncoding="utf-8"%>
<%@ taglib prefix="s" uri="/struts-tags" %>
<html>
  <head>
    <title>My JSP 'generatorTag.jsp' starting page</title>
  </head>
  <body>
    <h2>generator 标签示例</h2>
    generator 标签体内迭代子字符串集合:<br><br>
```

```
<!-- 使用 generator 标签将一个字符串解析成子字符串集合,并迭代输出集合 -->
<s:generator separator = ","
val = "'HTML 与 CSS 网页设计教学做一体化教程,JSP 网站设计教学做一体化教程,
软件工程教学做一体化教程,基于 Eclipse 平台的 JSP 应用教程,Struts 2 框架应用教程'">
<!-- 在 generator 标签体内,子字符串集合位于栈顶,所以这里的迭代就是临时生成的集合 -->
<s:iterator>
    <s:property/><br>
</s:iterator>
</s:generator>
<hr>
generator 标签体外迭代子字符串集合:<br><br>
<!-- 指定了 count,这里最多迭代 2 次 -->
<s:generator separator = "," count = "2" var = "books"
val = "'HTML 与 CSS 网页设计教学做一体化教程,JSP 网站设计教学做一体化教程,
软件工程教学做一体化教程,基于 Eclipse 平台的 JSP 应用教程,Struts 2 框架应用教程'">
</s:generator>
<!-- 在 generator 标签体外,这里的迭代是 PageContext 里的属性 books -->
<s:iterator value = "books" var = "book">
<s:property value = "book"/><br>
</s:iterator>
</body>
</html>
```

页面 generatorTag.jsp 的运行效果如图 6.12 所示。

### 5. merge 标签

merge 标签的用法与 append 标签完全一样,也是用于将多个集合拼接成一个集合,但它们的拼接方式不同。例如,需要拼接集合 List1 与 List2,使用 append 标签时,List2 追加到 List1 的尾部;而使用 merge 标签时,则先访问 List1 的第一个元素,再访问 List2 的第一个元素,然后才访问 List1 的第二个元素和 List2 的第二个元素,就这样一直交替访问。

merge 标签的使用示例与 append 标签的使用示例基本一样,此处不再赘述。

图 6.12 使用 generator 标签

### 6. subset 标签

subset 标签用于取得集合的子集,该标签的底层通过 org.apache.struts2.util.SubsetIteratorFilter 类提供实现。使用该标签时可指定如下几个属性。

(1) count:这是一个可选属性,该属性指定子集中元素的个数,如果不指定该属性,默认取得源集合的全部元素。

(2) source:这是一个可选属性,该属性指定源集合。如果不指定该属性,默认取得 ValueStack 栈顶的集合。

(3) start:这是一个可选属性,该属性指定子集从源集合的第几个元素开始截取。默认从第一个元素开始截取。

(4) decider:这是一个可选属性,该属性指定由开发者自己决定是否选中该元素。

Struts 2 允许开发者决定截取标准,如果开发者需要实现自己的截取标准,则需要编写一个实现 org.apache.struts2.util.SubsetIteratorFilter.Decider 接口的类,该类需要实现一个 public boolean decide(Object element)方法,如果该方法返回 true,则表明元素将被选入子集中。

下面通过一个简单示例讲解 subset 标签的使用方法。

首先,创建一个实现 Decider 接口的类,具体代码如下:

```java
package bean;
import org.apache.struts2.util.SubsetIteratorFilter.Decider;
public class MyDecider implements Decider{
    @Override
    public boolean decide(Object element) throws Exception {
        String s = (String)element;
        //如果集合元素中包含"应用教程"字串,即可入选子集
        return s.contains("应用教程");
    }
}
```

其次,创建 JSP 页面 subsetTag.jsp,具体代码如下:

```jsp
<%@ page language="java" import="java.util.*" pageEncoding="utf-8"%>
<%@taglib prefix="s" uri="/struts-tags" %>
<html>
  <head>
    <title>My JSP 'subsetTag.jsp' starting page</title>
  </head>
  <body>
    <h3>subset 标签示例</h3>
    <!-- 定义一个 Decider 实例 -->
    <s:bean name="bean.MyDecider" var="mydecider"/>
    <!-- 使用自定义的 Decider 实例来截取目标集合,生成子集 -->
    <s:subset decider="#mydecider"
    source="{'HTML 与 CSS 网页设计教学做一体化教程','JSP 网站设计教学做一体化教程','软件工程教学做一体化教程','基于 Eclipse 平台的 JSP 应用教程','Struts 2 框架应用教程'}">
    <!-- 迭代 subset 标签实现的子集 -->
    <s:iterator>
    <s:property/><br>
    </s:iterator>
    </s:subset>
  </body>
</html>
```

页面 subsetTag.jsp 的运行效果如图 6.13 所示。

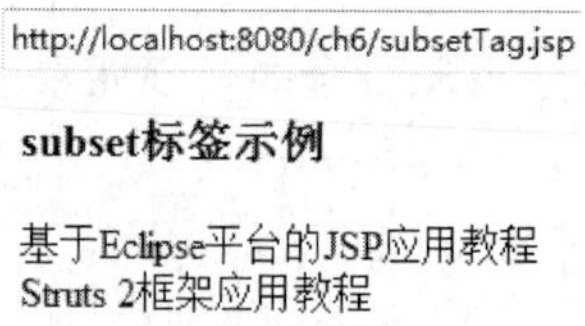

### 7. sort 标签

sort 标签用于对指定的集合元素进行排序,进行排序时,必须提供自己的排序规则,即实现自己的 Comparator 类,该

图 6.13 使用 subset 标签

类需要实现 java.util.Comparator 接口。使用 sort 标签时可指定如下属性。

（1）comparator：这是一个必填属性，该属性指定进行排序的 Comparator 实例。

（2）source：这是一个可选的属性，该属性指定被排序的集合。如果不指定该属性，则对 ValueStack 栈顶的集合进行排序。

需要注意的是，在 sort 标签体内，sort 标签排序后的集合放在 ValueStack 的栈顶，如果该标签结束后，则排序后的集合将移出值栈。

下面通过一个简单示例讲解 sort 标签的使用方法。

首先，创建自己的 Comparator 类，具体代码如下：

```java
package bean;
import java.util.Comparator;
public class MyComparator implements Comparator<String>{
    //决定排序规则
    @Override
    public int compare(String element1, String element2) {
        //元素按照其长度从小到大进行排序
        return element1.length() - element2.length();
    }
}
```

其次，创建 JSP 页面 sortTag.jsp，具体代码如下：

```jsp
<%@ page language="java" import="java.util.*" pageEncoding="utf-8"%>
<%@taglib prefix="s" uri="/struts-tags" %>
<html>
  <head>
    <title>My JSP 'sortTag.jsp' starting page</title>
  </head>
  <body>
    <h3>sort 标签示例</h3>
    <!-- 定义一个 Comparator 实例 -->
    <s:bean name="bean.MyComparator" var="myComparator"/>
    <!-- 使用自定义的排序规则,对目标集合进行排序 -->
    <s:sort comparator="#myComparator" source="{'HTML 与 CSS 网页设计教学做一体化教程','JSP 网站设计教学做一体化教程','软件工程教学做一体化教程','基于 Eclipse 平台的 JSP 应用教程','Struts 2 框架应用教程'}">
      <!-- 迭代输出排序后的集合 -->
      <s:iterator>
        <s:property/><br >
      </s:iterator>
    </s:sort>
  </body>
</html>
```

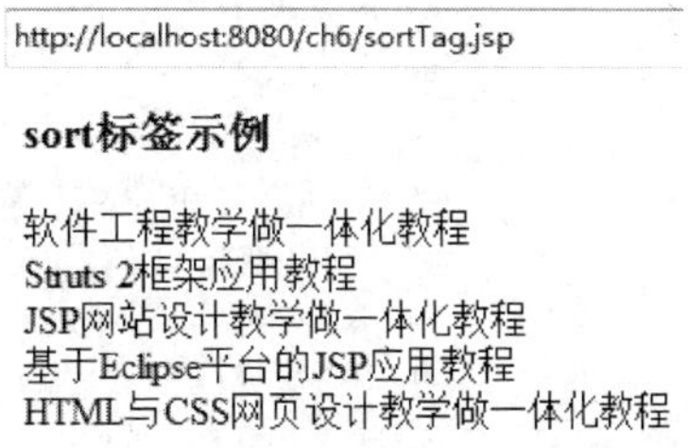

页面 sortTag.jsp 的运行效果如图 6.14 所示。

图 6.14　使用 sort 标签

### 6.2.2 能力目标

掌握常用的流程控制标签的使用方法。

### 6.2.3 任务驱动

1) 任务的主要内容

首先，编写一个 Action 类 Task2Action，在 execute 方法中创建 User 类(6.1.1节)的多个对象，并将这些对象保存在 ArrayList＜User＞集合中；同时，将 ArrayList＜User＞集合对象存储在 request 中。

其次，编写一个 JSP 页面 task_6_2.jsp(Action 成功跳转页面)，在该 JSP 页面中使用 iterator 标签迭代输出 ArrayList＜User＞集合中的数据。

最后，通过"http://localhost:8080/ch6/task2.action"访问 Action。

2) 任务的代码模板

Task2Action.java 的代码模板如下：

```java
package action;
import java.util.ArrayList;
import java.util.Map;
import org.apache.struts2.interceptor.RequestAware;
import bean.User;
import com.opensymphony.xwork2.ActionSupport;
public class Task2Action extends ActionSupport implements RequestAware{
    private static final long serialVersionUID = 1L;
    Map<String, Object> request;
    public String execute(){
        ArrayList<User> arr = new ArrayList<User>();
        //创建 User 的对象
        User u1 = new User();
        u1.setUname("张三");
        u1.setUemail("zhangsan@edu.cn");
        //将 User 的对象保存到 ArrayList 集合中
        arr.add(u1);

        User u2 = new User();
        u2.setUname("李四");
        u2.setUemail("lisi@edu.cn");
        arr.add(u2);

        User u3 = new User();
        u3.setUname("王五");
        u3.setUemail("wangwu@edu.cn");
        arr.add(u3);

        //代码1 将 ArrayList<User>集合对象 arr 以"myUsers"为关键字保存在 request 中
        【代码1】
        return SUCCESS;
    }
}
```

```
        @Override
        public void setRequest(Map<String, Object> arg0) {
            request = arg0;
        }
    }
```

task_6_2.jsp 的代码模板如下:

```
<%@ page language="java" import="java.util.*" pageEncoding="UTF-8"%>
<%@taglib prefix="s" uri="/struts-tags" %>
<%
String path = request.getContextPath();
String basePath = request.getScheme()+"://"+request.getServerName()+":"+request.getServerPort()+path+"/";
%>
<!DOCTYPE HTML PUBLIC "-//W3C//DTD HTML 4.01 Transitional//EN">
<html>
  <head>
    <base href="<%=basePath%>">
    <title>My JSP 'task_6_2.jsp' starting page</title>
  </head>
  <body>
    使用 iterator 标签迭代输出 ArrayList 集合数据:<br>
    <!-- 代码 2,取出 Action 中 request 的数据 -->
    <s:iterator var="u" value="【代码 2】">
        <!-- 代码 3,使用 property 标签输出 uname 属性值 -->
        【代码 3】  
        <!-- 代码 4,使用 property 标签输出 uemail 属性值 -->
        【代码 4】<br>
    </s:iterator>
  </body>
</html>
```

Action 的配置代码略。

3) 任务小结或知识扩展

任务中 ArrayList<User>集合还可以通过 Action 类的属性(需要 get 和 set 方法)传递到页面显示。具体代码如下:

Action 类的代码如下:

```
package action;
import java.util.ArrayList;
import bean.User;
import com.opensymphony.xwork2.ActionSupport;
public class Task2Action extends ActionSupport {
    private static final long serialVersionUID = 1L;
    ArrayList<User> myUsers;
    public String execute(){
        myUsers = new ArrayList<User>();
        //创建 User 的对象
        User u1 = new User();
```

```java
        u1.setUname("张三");
        u1.setUemail("zhangsan@edu.cn");
        //将 User 的对象保存到 ArrayList 集合中
        myUsers.add(u1);

        User u2 = new User();
        u2.setUname("李四");
        u2.setUemail("lisi@edu.cn");
        myUsers.add(u2);

        User u3 = new User();
        u3.setUname("王五");
        u3.setUemail("wangwu@edu.cn");
        myUsers.add(u3);

        return SUCCESS;
    }
    public ArrayList<User> getMyUsers() {
        return myUsers;
    }
    public void setMyUsers(ArrayList<User> myUsers) {
        this.myUsers = myUsers;
    }
}
```

### JSP 页面的代码如下：

```jsp
<%@ page language="java" import="java.util.*" pageEncoding="UTF-8"%>
<%@taglib prefix="s" uri="/struts-tags" %>
<%
String path = request.getContextPath();
String basePath = request.getScheme()+"://"+request.getServerName()+":"+request.getServerPort()+path+"/";
%>
<!DOCTYPE HTML PUBLIC "-//W3C//DTD HTML 4.01 Transitional//EN">
<html>
  <head>
    <base href="<%=basePath%>">
    <title>My JSP 'task_6_2.jsp' starting page</title>
  </head>
  <body>
    使用 iterator 标签迭代输出 ArrayList 集合数据：<br>
    <!-- 取出 Action 中 myUsers 属性值 -->
    <s:iterator var="u" value="myUsers">
    <!-- 使用 property 标签输出 uname 属性值 -->
    <s:property value="uname"/>  
    <!-- 使用 property 标签输出 uemail 属性值 -->
    <s:property value="uemail"/><br>
    </s:iterator>
  </body>
</html>
```

（4）任务代码模板的参考答案

【代码1】request.put("myUsers", arr);

【代码2】♯request.myUsers

【代码3】<s:property value = "uname"/>

【代码4】<s:property value = "uemail"/>

### 6.2.4 实践环节

（1）创建JSP页面practice624_1.jsp，在该页面中分别使用append标签和merge标签对集合：{'我','爱','学','习'}和{'Java Web开发','以及','Struts 2框架','。'}进行拼接，并使用iterator标签输出拼接后的集合，试分析append标签和merge标签的不同。页面practice624_1.jsp的运行效果如图6.15所示。

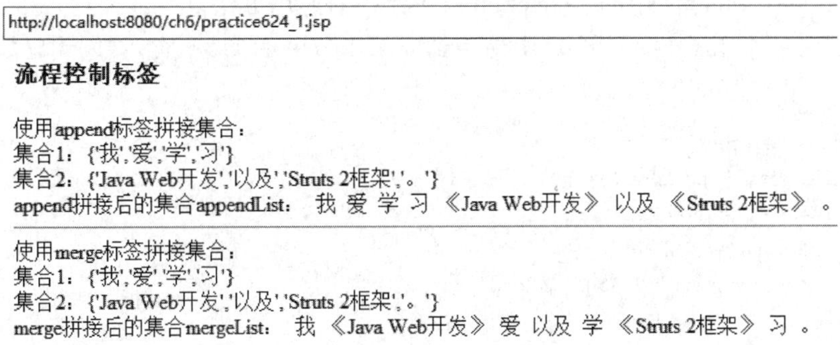

```
http://localhost:8080/ch6/practice624_1.jsp

流程控制标签

使用append标签拼接集合：
集合1：{'我','爱','学','习'}
集合2：{'Java Web开发','以及','Struts 2框架','。'}
append拼接后的集合appendList：  我 爱 学 习 《Java Web开发》 以及 《Struts 2框架》 。

使用merge标签拼接集合：
集合1：{'我','爱','学','习'}
集合2：{'Java Web开发','以及','Struts 2框架','。'}
merge拼接后的集合mergeList：  我 《Java Web开发》 爱 以及 学 《Struts 2框架》 习 。
```

图6.15　practice624_1.jsp的运行效果

（2）创建JSP页面practice624_2.jsp，在该页面中使用subset标签截取集合♯{'Java':'5学分','C':'4学分','C++':'6学分','JSP':'5学分','Struts 2':'5学分'}，指定截取位置为1，截取3个元素。并使用iterator标签输出子集下标为奇数的元素。页面practice624_2.jsp的运行效果如图6.16所示。

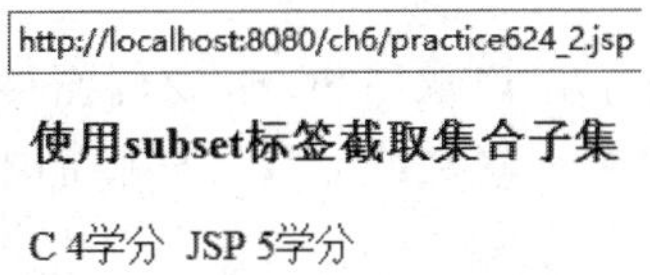

```
http://localhost:8080/ch6/practice624_2.jsp

使用subset标签截取集合子集

C 4学分  JSP 5学分
```

图6.16　practice624_2.jsp的运行效果

## 6.3　UI标签

UI标签，主要用来生成HTML元素的标签，又分为表单标签和非表单标签。Struts 2的表单标签是用户最常用的标签，这些表单标签都包含了非常多的属性，但有很多属性完全是通用的。大部分表单标签和HTML表单元素之间一一对应，只要读者熟悉HTML中表单的相关元素，则学习Struts 2的表单标签是比较轻松的。本节重点介绍几个与HTML元素差别较大的标签，那些相似的标签（如文本、按钮等），读者可自主学习。

### 6.3.1 核心知识

**1. checkboxlist 标签**

checkboxlist 标签可以一次创建多个复选框,用于一次生成多个 HTML 标签中的 <input type="checkbox".../>,它根据 list 属性指定的集合生成多个复选框。因此,该标签有个必填属性 list,其他属性大部分都是通用属性。另外,checkboxlist 标签还有两个常用属性。

(1) listKey:指定集合元素中的某个属性作为复选框的 value。如果集合为 Map 类型则可以使用 key 和 value 分别代表 Map 对象的 key 和 value 作为复选框的 value。

(2) listValue:指定集合元素中的某个属性作为复选框的 label。如果集合为 Map 类型则可以使用 key 和 value 分别代表 Map 对象的 key 和 value 作为复选框的 label。

下面通过一个简单示例讲解 checkboxlist 标签的使用方法。

首先,定义一个 JavaBean 类,其中使用到 6.1.1 节中的 User 类,具体代码如下:

```java
package bean;
public class MyUser {
    public User[] getUsers(){
        User u1 = new User();
        u1.setUname("陈恒");
        u1.setUemail("123456@qq.com");
        User u2 = new User();
        u2.setUname("张一鸣");
        u2.setUemail("888888@qq.com");
        User[] users = new User[2];
        users[0] = u1;
        users[1] = u2;
        return users;
    }
}
```

其次,创建 JSP 页面 checkboxlistTag.jsp,在该页面中分别使用了简单集合、Map 对象、集合里存放 Java 实例来创建多个复选框。具体代码如下:

```jsp
<%@ page language="java" import="java.util.*" pageEncoding="utf-8"%>
<%@taglib prefix="s" uri="/struts-tags" %>
<html>
  <head>
    <title>My JSP 'checkboxlistTag.jsp' starting page</title>
  </head>
  <body>
    <h3>checkboxlist 标签示例</h3>
    <s:form>
    <!-- 使用简单集合生成多个复选框 -->
    <s:checkboxlist name="mysongers" label="请选择您喜欢的歌手" labelposition="top"
       list="{'张三','李四','王五','张四','李五'}"/>
    <!-- 使用 Map 对象生成多个复选框 -->
```

```
    <s:checkboxlist name = "mysports" label = "请选择您喜欢的运动" labelposition = "top"
    list = "#{1:'瑜伽用品',2:'户外用品',3:'球类',4:'自行车'}" listKey = "key"
    listValue = "value" value = "{1,2,3}"/>
    <!-- 创建 MyUser 实例 -->
    <s:bean name = "bean.MyUser" var = "mu"/>
    <!-- 使用集合里的实例来生成多个复选框 -->
    <s:checkboxlist name = "myusers" label = "请选择您喜欢的用户" labelposition = "top"
    list = "#mu.users" listKey = "uemail" listValue = "uname" />
    <!-- #mu.users 相当于对象 mu 调用 getUsers()方法 -->
    <!-- listKey = "uemail"是指将集合中实例的 uemail 作为复选框的 value -->
    <!-- listKey = "uname"是指将集合中实例的 uname 作为复选框的 Label,即显示在页面上 -->
    </s:form>
  </body>
</html>
```

页面 checkboxlistTag.jsp 的运行效果如图 6.17 所示。

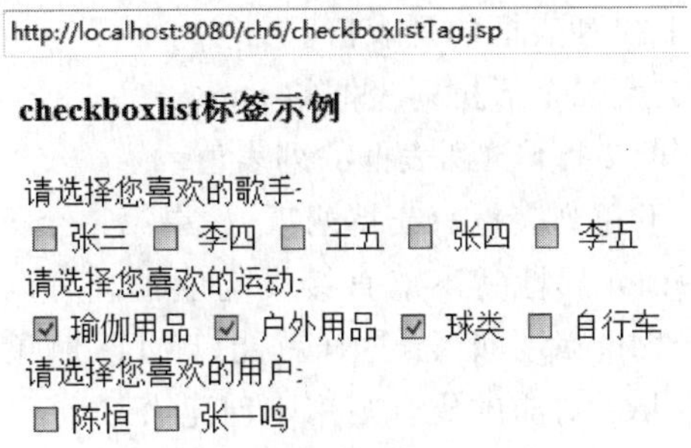

图 6.17　使用 checkboxlist 标签

### 2. combobox 标签

combobox 标签生成一个单行文本框和下拉列表框的组合,但两个表单元素只对应一个请求参数,只有单行文本框里的值才包含请求参数,而下拉列表框只是用于辅助输入,没有产生请求参数。使用该标签时,需要指定一个 list 属性,该属性指定的集合将用于生成列表项。

下面通过一个简单示例讲解 combobox 标签的使用方法。

首先,创建 JSP 页面 comboboxTag.jsp,具体代码如下:

```
<%@ page language = "java" import = "java.util.*" pageEncoding = "utf-8" %>
<%@ taglib prefix = "s" uri = "/struts-tags" %>
<html>
  <head>
    <title>My JSP 'comboboxTag.jsp' starting page</title>
  </head>
  <body>
    <s:form>
    <h3>combobox 标签示例</h3>
    <!-- 使用 combobox 标签,其中 list 指定下拉列表选项 -->
    <s:combobox label = "请选择您喜欢的大学" size = "20" maxlength = "20" name = "collage"
    list = "{'清华大学','北京大学','南京大学','科技大学'}"/>
```

```
        </s:form>
    </body>
</html>
```

其次,运行 comboboxTag.jsp 页面,效果如图 6.18 所示。

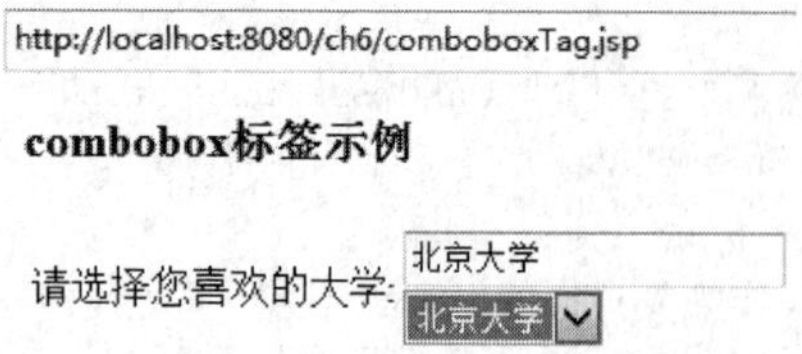

图 6.18 使用 combobox 标签

**注意**:combobox 标签与 select 标签不同的是,combobox 标签无须指定 listKey 和 listValue 属性,因为 combobox 标签只是用于辅助输入,而不用于发送请求参数。

### 3. doubleselect 标签

doubleselect 标签会生成一个级联列表框,当选择第一个下拉列表时,第二个下拉列表框的内容会随之改变。doubleselect 标签的常用属性如下。

(1) name:指定第一个下拉列表框的 name。
(2) list:指定第一个下拉列表框的列表值。
(3) doubleName:指定第二个下拉列表框的 name。
(4) doubleList:指定第二个下拉列表框的列表值。

在默认情况下,第一个下拉列表框只支持两项,如果第一个下拉列表框包含 3 个或更多的值,这里的 list 和 doubleList 属性就不能直接设定值了。可以这样实现:首先定义一个 Map 对象,该 Map 对象的 value 都是集合,这样就能以 Map 对象的多个 key 创建第一个下拉列表框的列表项,而每个 key 对应的集合则用于创建第二个下拉列表框的列表项。

下面通过一个示例讲解 doubleselect 标签的使用方法。

首先,创建一个 JSP 页面 doubleselectTag.jsp,具体代码如下:

```
<%@ page language="java" import="java.util.*" pageEncoding="utf-8"%>
<%@ taglib prefix="s" uri="/struts-tags" %>
<html>
    <head>
        <title>My JSP 'doubleselectTag.jsp' starting page</title>
    </head>
    <body>
        <h3>doubleselect 标签示例</h3>
        <s:form theme="simple">
        <table><tr>
            <td>选择您去过的城市:</td>
            <td><s:doubleselect name="country" list="{'国内','国外'}"
            doubleList="top=='国内'?{'北京','上海','广州','深圳','大连'}:{'纽约','伦敦','东京'}"
            doubleName="city"/>
            <!-- 使用{'国内','国外'}作为第一个下拉列表的选项 -->
            <!-- 第二个下拉列表,根据前一个选项来确定值,doubleList 的值是一个三目运算符表达式,top 代表第一个下拉列表选中的值,意思是当第一个下拉列表选择"国内"时,第二列表就使用{'北京','上海','广州','深圳','大连'}来创建,否则使用{'纽约','伦敦','东京'}来创建 -->
            </td>
```

```
        <td>选择您的学校: </td>
        <s:set var = "myschools" value = "#{'小学':{'东师附小','北师附小','科大附小'},'中学':
{'北京四中','上海一中','深圳八中','大连二十四中'},'大学':{'北京大学','清华大学','中国科
大'}}"/>
        <td><s:doubleselect name = "bank" list = "#myschools.keySet()"
            doubleList = "#myschools[top]"
            doubleName = "school"/>
        </td>
      </tr></table>
    </s:form>
  </body>
</html>
```

其次,运行 doubleselectTag.jsp 页面,效果如图 6.19 所示。

图 6.19  使用 doubleselect 标签

### 4. optiontransferselect 标签

optiontransferselect 是一个比较复杂的标签,它会创建两个选项用来转移下拉列表项,该标签会生成两个<select.../>标签,并且会生成一系列的按钮,这些按钮可以控制选项在两个下拉列表之间移动、升降。当提交该表单时,两个<select.../>标签的请求参数都会被提交。optiontransferselect 标签具有很多属性,常用的属性如下。

(1) addAllToLeftLabel:设置全部移动到左边按钮上的文本。

(2) addAllToRightLabel:设置全部移动到右边按钮上的文本。

(3) addToLeftLabel:设置向左移动按钮上的文本。

(4) addToRightLabel:设置向右移动按钮上的文本。

(5) allowAddAllToLeft:设置是否出现全部移动到左边的按钮,默认值为 true。

(6) allowAddAllToRight:设置是否出现全部移动到右边的按钮,默认值为 true。

(7) allowAddToLeft:设置是否出现移动到左边的按钮,默认值为 true。

(8) allowAddToRight:设置是否出现移动到右边的按钮,默认值为 true。

(9) allowSelectAll:设置是否出现全部选择的按钮,默认值为 true。

(10) doubleList:这是必选属性,设置右边下拉列表框的集合。

(11) doubleListKey:设置右边下拉列表框的选项 value 属性。

(12) doubleListValue:设置右边下拉列表框的选项 label 属性。

（13）doubleName：这是必选属性，设置右边下拉列表框的 name 属性。

（14）doubleValue：设置右边下拉列表框的 value 属性。

（15）doubleMultiple：设置右边下拉列表框是否允许多选，默认值为 false。

（16）doubleCssStyle：设置右边下拉列表框的样式。

（17）leftTitle：设置左边列表的标题。

（18）rightTitle：设置右边列表的标题。

（19）leftUpLabel：设置左边上移按钮上的文本。

（20）leftDownLabel：设置左边下移按钮上的文本。

（21）rightUpLabel：设置右边上移按钮上的文本。

（22）rightDownLabel：设置右边下移按钮上的文本。

（23）list：这是必选属性，设置左边下拉列表框的集合。

（24）listKey：设置左边下拉列表框的选项 value 属性。

（25）listValue：设置左边下拉列表框的选项 label 属性。

（26）multiple：设置左边下拉列表框是否允许多选，默认值为 false。

（27）name：这是必选属性，设置左边下拉列表框的 name 属性。

（28）value：设置左边下拉列表框的 value 属性。

虽然 optiontransferselect 标签具有很多属性，但一般不会用到全部属性，下面通过一个示例来学习 optiontransferselect 标签的使用方法。

首先，创建一个 JSP 页面 optiontransferselectTag.jsp，具体代码如下：

```jsp
<%@ page language="java" import="java.util.*" pageEncoding="utf-8" %>
<%@ taglib prefix="s" uri="/struts-tags" %>
<html>
  <head>
    <title>My JSP 'optiontransferselectTag.jsp' starting page</title>
  </head>
  <body>
    <s:form>
        <s:optiontransferselect
            label="选择您喜欢的课程"
            name="left"
            doubleName="right"
            leftTitle="计算机课程"
            rightTitle="非计算机课程"
            leftUpLabel="上移"
            leftDownLabel="下移"
            rightUpLabel="上移"
            rightDownLabel="下移"
            addToLeftLabel="<-向左移动"
            addToRightLabel="向右移动->"
            addAllToLeftLabel="<-全部左移"
            addAllToRightLabel="全部右移->"
            selectAllLabel="-全部选择-"
            cssStyle="width:150x;height:200px;"
            doubleCssStyle="width:150x;height:200px;"
```

```
                list = "♯{1:'Java 程序设计',2:'C++程序设计',3:'操作系统',4:'高等数学 I',5:'海峡
        两岸'}"
                listKey = "key"
                listValue = "value"
                multiple = "true"
                doubleList = "♯{6:'大学物理',7:'大学化学',8:'计算机英语',9:'JSP 程序设计'}"
                doubleListKey = "key"
                doubleListValue = "value"
                doubleMultiple = "true"/>
        </s:form >
    </body >
</html>
```

其次,运行 optiontransferselectTag.jsp 页面,效果如图 6.20 所示。

图 6.20 使用 optiontransferselect 标签

### 5. select 标签

select 标签用于生成一个下拉列表,通过为该标签指定 list 属性,系统会使用 list 属性指定的集合(List、Map 或集合元素是对象的集合)生成下拉列表的选项。select 标签的常用属性如下。

(1) listKey:该属性指定集合元素中的某个属性(例如集合元素为 User 实例,指定 User 实例的 uname 属性)作为下拉列表选项的值(value)。如果集合是 Map,则可以使用 key 和 value 值分别代表 Map 对象的关键字和值作为下拉列表选项的值,即:listKey = "key"代表使用 Map 对象的关键字作为下拉列表选项的值;listKey = "value"代表使用 Map 对象的值作为下拉列表选项的值。

(2) listValue:该属性指定集合元素中的某个属性(例如集合元素为 User 实例,指定 User 实例的 uname 属性)作为下拉列表选项的标签。如果集合是 Map,则可以使用 key 和 value 值分别代表 Map 对象的关键字和值作为下拉列表选项的标签,即:listValue = "key"代表使用 Map 对象的关键字作为下拉列表选项的标签;listValue = "value"代表使用 Map 对象的值作为下拉列表选项的标签。

(3) multiple:设置是否允许多选,默认为 false。

下面通过一个简单示例讲解 select 标签的使用方法。

首先,创建 JSP 页面 selectTag.jsp,在该页面中分别使用 List、Map、集合里放置

JavaBean 实例来创建多个下拉列表。具体代码如下：

```jsp
<%@ page language="java" import="java.util.*" pageEncoding="utf-8"%>
<%@taglib prefix="s" uri="/struts-tags" %>
<html>
  <head>
    <title>My JSP 'selectTag.jsp' starting page</title>
  </head>
  <body>
    <h3>select 标签示例</h3>
    <s:form>
    <!-- 使用 List 生成下拉列表,value 属性指定默认选择项 -->
    <s:select name="course" label="选择您喜欢的课程"
      list="{'Java','C','.Net','JSP','Struts 2'}" value="'JSP'"/>
    <!-- 使用 Map 生成下拉列表,value 属性指定默认选择项 -->
    <s:select name="sport" label="选择您喜欢的运动"
      list="#{1:'瑜伽',2:'户外',3:'球类',4:'自行车'}"
      listKey="key" listValue="value" value="2"/>
    <!-- 集合里放置多个 JavaBean 实例生成下拉列表 -->
    <!-- 创建 MyUser(checkboxlist 标签中的 MyUser 类)实例 -->
    <s:bean name="bean.MyUser" var="mu"/>
    <s:select name="yourUser" list="#mu.users" label="选择您喜欢的用户"
      listKey="uemail" listValue="uname"/>
    <!-- #mu.users 相当于对象 mu 调用 getUsers()方法 -->
    <!-- listKey="uemail"是指将集合中实例的 uemail 作为选项的 value -->
    <!-- listKey="uname"是指将集合中实例的 uname 作为选项框的 label,即显示在页面上 -->
    </s:form>
  </body>
</html>
```

其次,运行 selectTag.jsp 页面,效果如图 6.21 所示。

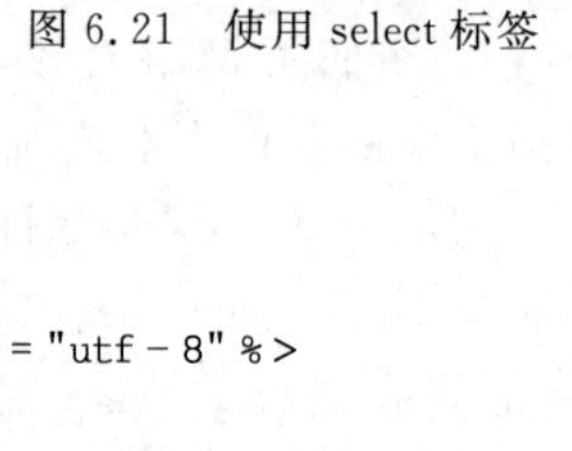

#### 6. optgroup 标签

optgroup 标签用于生成一个下拉列表的选项组,该标签必须放在 select 标签体内使用。使用 optgroup 标签时,与使用 select 标签类似,一样需要指定 list、listKey 和 listValue 等属性,而且属性的含义也与 select 标签的属性含义相同。还可以通过 label 属性为选项组指定组名。

图 6.21 使用 select 标签

下面通过一个示例讲解 optgroup 标签的使用方法。

首先,创建一个 JSP 页面 optgroupTag.jsp,具体代码如下：

```jsp
<%@ page language="java" import="java.util.*" pageEncoding="utf-8"%>
<%@taglib prefix="s" uri="/struts-tags" %>
<html>
  <head>
    <title>My JSP 'optgroupTag.jsp' starting page</title>
  </head>
  <body>
    <h3>optgroup 标签示例</h3>
```

```
<s:form>
    <s:select name = "course" label = "选择您喜欢的技术" multiple = "true"
    size = "10" list = "{'C','.Net','PHP'}">
        <s:optgroup label = "Java 系列"
        list = "#{1:'JSP',2:'Struts 2',3:'Hibernate',4:'Spring'}"
        listKey = "key" listValue = "value"/>
        <s:optgroup label = "其他"
        list = "#{1:'JSF',2:'CSS',3:'HTML'}"
        listKey = "key" listValue = "value"/>
    </s:select>
</s:form>
</body>
</html>
```

其次,运行 optgroupTag.jsp 页面,效果如图 6.22 所示。

### 7. radio 标签

radio 标签的用法与 checkboxlist 的用法几乎完全相同,唯一不同的是 checkboxlist 生成的是复选框,而 radio 生成的是单选按钮。

图 6.22 使用 optgroup 标签

下面通过一个简单示例讲解 radio 标签的使用方法。

首先,创建 JSP 页面 radioTag.jsp,具体代码如下:

```
<%@ page language = "java" import = "java.util.*" pageEncoding = "utf-8"%>
<%@ taglib prefix = "s" uri = "/struts-tags" %>
<html>
  <head>
    <title>My JSP 'radioTag.jsp' starting page</title>
  </head>
  <body>
    <h3>radio 标签示例</h3>
    <s:form>
    <!-- 使用简单集合生成多个单选按钮 -->
    <s:radio name = "mysongers" label = "请选择您喜欢的歌手"
    list = "{'张三','李四','王五','张四','李五'}"/>
    <!-- 使用 Map 对象生成多个单选按钮 -->
    <s:radio name = "mysports" label = "请选择您喜欢的运动"
    list = "#{1:'瑜伽用品',2:'户外用品',3:'球类',4:'自行车'}" listKey = "key"
    listValue = "value" />
    <!-- 创建 MyUser(checkboxlist 标签中的类)实例 -->
    <s:bean name = "bean.MyUser" var = "mu"/>
    <!-- 使用集合里的实例来生成多个单选按钮 -->
    <s:radio name = "myusers" label = "请选择您喜欢的用户"
    list = "#mu.users" listKey = "uemail" listValue = "uname" />
    <!-- #mu.users 相当于对象 mu 调用 getUsers()方法 -->
    <!-- listKey = "uemail"是指将集合中实例的 uemail 作为单选按钮的 value -->
    <!-- listKey = "uname"是指将集合中实例的 uname 作为单选按钮的 label,即显示在页面上 -->
```

```
        </s:form>
    </body>
</html>
```

其次,运行 radioTag.jsp 页面,效果如图 6.23 所示。

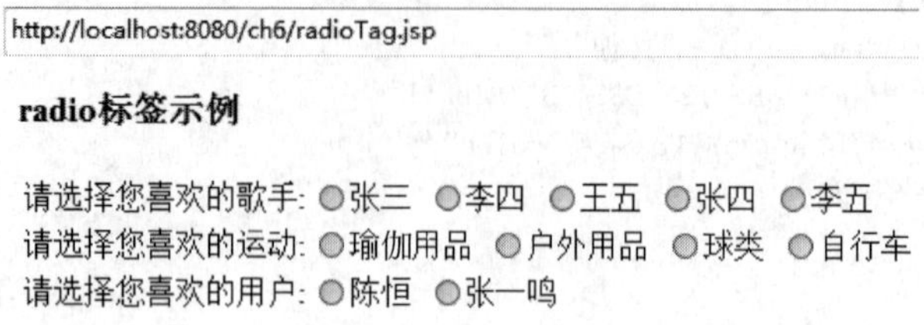

图 6.23 使用 radio 标签

### 6.3.2 能力目标

掌握常用的表单标签的使用方法。

### 6.3.3 任务驱动

1) 任务的主要内容

首先,编写一个 Action 类 Task3Action,该 Action 类有一个 ArrayList<User>类型的属性 myUsers;在 Action 类的 execute 方法中创建多个 User 类(6.1.1 节)的对象,并将这些对象存储到 myUsers 集合中。

图 6.24 使用 ArrayList<User>集合生成下拉列表

其次,编写一个 JSP 页面 task_6_3.jsp(Action 成功跳转页面),在该页面中使用 Task3Action 类的属性 myUsers 生成一个下拉列表。User 类的 uemail 属性作为下拉列表选项的值;User 类的 uname 属性作为下拉列表选项的文本。

最后,通过 "http://localhost:8080/ch6/task3.action"访问 Action。运行效果如图 6.24 所示。

2) 任务的代码模板

Task3Action.java 的代码如下:

```
package action;
import java.util.ArrayList;
import bean.User;
import com.opensymphony.xwork2.ActionSupport;
public class Task3Action extends ActionSupport{
    private static final long serialVersionUID = 1L;
    private ArrayList<User> myUsers;
    public String execute(){
        myUsers = new ArrayList<User>();
        //创建 User 的对象
        User u1 = new User();
        u1.setUname("张三");
```

```java
        u1.setUemail("zhangsan@edu.cn");
        //将 User 的对象保存到 ArrayList 集合中
        myUsers.add(u1);

        User u2 = new User();
        u2.setUname("李四");
        u2.setUemail("lisi@edu.cn");
        myUsers.add(u2);

        User u3 = new User();
        u3.setUname("王五");
        u3.setUemail("wangwu@edu.cn");
        myUsers.add(u3);

        return SUCCESS;
    }
    public ArrayList<User> getMyUsers() {
        return myUsers;
    }
    public void setMyUsers(ArrayList<User> myUsers) {
        this.myUsers = myUsers;
    }
}
```

task_6_3.jsp 的代码模板如下：

```jsp
<%@ page language="java" import="java.util.*" pageEncoding="UTF-8"%>
<%@taglib prefix="s" uri="/struts-tags" %>
<%
    String path = request.getContextPath();
    String basePath = request.getScheme()+"://"+request.getServerName()+":"+request.
    getServerPort()+path+"/";
%>
<!DOCTYPE HTML PUBLIC "-//W3C//DTD HTML 4.01 Transitional//EN">
<html>
  <head>
    <base href="<%=basePath%>">
    <title>My JSP 'task_6_3.jsp' starting page</title>
  </head>
  <body>
    <s:form theme="simple">
    选择您的用户：
    <!--
    代码 1 将 Action 类的属性 myUsers 作为下拉列表的选项；
    代码 2 将集合中实例的 uemail 作为下拉列表选项的 value；
    代码 3 将集合中实例的 uname 作为下拉列表选项的文本显示在页面 -->
    <s:select name="myuser"【代码 1】【代码 2】【代码 3】/>
    </s:form>
  </body>
</html>
```

### 3) 任务小结或知识扩展

UI 标签除了表单标签外，还有 actionerror、actionmessage 以及 fielderror 等非表单标签。

（1）actionerror：如果 Action 实例的 getActionErrors() 方法返回不为 null，则该标签负责输出该方法返回的系列错误。

（2）actionmessage：如果 Action 实例的 getActionMessages() 方法返回不为 null，则该标签负责输出该方法返回的系列消息。

（3）fielderror：如果 Action 实例存在表单域的类型转换错误、校验错误，则该标签负责输出这些错误提示。该标签的使用细节将在第 7 章 Struts 2 的输入校验进行讲解。

actionerror 和 actionmessage 标签主要应用在业务逻辑错误消息的提示，而 fielderror 标签主要应用在输入校验错误消息的提示。

下面通过一个简单的登录示例讲解 actionerror 和 actionmessage 标签的使用方法。

首先，创建一个 JSP 页面 login.jsp，具体代码如下：

```jsp
<%@ page language="java" import="java.util.*" pageEncoding="utf-8"%>
<%@ taglib prefix="s" uri="/struts-tags" %>
<%
    String path = request.getContextPath();
    String basePath = request.getScheme()+"://"+request.getServerName()+":"+request.getServerPort()+path+"/";
%>
<html>
  <head>
    <base href="<%=basePath%>">
    <title>My JSP 'login.jsp' starting page</title>
  </head>
  <body>
    <s:form action="login.action" method="post" theme="simple">
    <table>
    <tr>
        <td>uname:</td>
        <td><s:textfield name="uname"/></td>
    </tr>
    <tr>
        <td>upass:</td>
        <td><s:password name="upass"/></td>
    </tr>
    <tr>
        <td colspan="2" align="center"><s:submit value="提交"/></td>
    </tr>
    </table>
    <s:actionmessage/>
    <s:actionerror/>
    </s:form>
  </body>
</html>
```

其次，创建登录业务处理的 Action 类 LoginAction.java，具体代码如下：

```java
package action;
import com.opensymphony.xwork2.ActionSupport;
public class LoginAction extends ActionSupport{
    private String uname;
    private String upass;
    public String getUname() {
        return uname;
    }
    public void setUname(String uname) {
        this.uname = uname;
    }
    public String getUpass() {
        return upass;
    }
    public void setUpass(String upass) {
        this.upass = upass;
    }
    public String execute(){
        if(!"chenheng".equals(uname)){
            //添加业务逻辑错误消息
            this.addActionMessage("用户名错误!");
            return "fail";
        }else if(!"123456".equals(upass)){
            //添加业务逻辑错误消息
            this.addActionError("密码错误!");
            return "fail";
        }
        return SUCCESS;
    }
}
```

再次，配置 Action，具体代码如下：

```
<action name="login" class="action.LoginAction">
    <result>/index.jsp</result>
    <result name="fail">/login.jsp</result>
</action>
```

最后，运行 login.jsp，分别输入错误的用户名和密码进行测试，运行效果如图 6.25 所示。

4）任务代码模板的参考答案

【代码 1】list = "myUsers"

【代码 2】listKey = "uemail"

【代码 3】listValue = "uname"

图 6.25 使用 actionerror 和 actionmessage 标签

## 6.3.4 实践环节

编写一个 JSP 页面 practice634.jsp,在该页面中使用 Struts 2 的表单标签完成 UI 设计,页面运行效果如图 6.26 所示。具体要求如下。

(1) 使用 Map 集合对象创建"性别"单选按钮,默认选项为"女"。

(2) 使用 Map 集合对象创建"所在地"下拉列表,默认选项为"中国"。

(3) 使用 List 集合对象创建"选择您去过的城市"级联列表框,第一个下拉列表只有两个选项:"国内"和"国外"。

(4) 使用 List 集合对象创建"喜欢的歌手"复选框。

图 6.26　Struts 2 表单标签

## 6.4　本章小结

本章从标签库的分类讲起,介绍了标签库的用法和用处。本章的重点是介绍 Struts 2 标签库的用法。详细讲解了 Struts 2 的流程控制标签、数据标签、表单标签、非表单标签各个属性的意义,并且使用丰富的示例代码示范了这些标签的用法。

## 习　题　6

1. 不属于 Struts 2 标签的是(　　)。
　　A.＜s:textfield＞　　B.＜s:textarea＞　　C.＜s:submit＞　　D.＜select＞
2. 在 JSP 页面中使用 Struts 2 标签时,需要事先导入(　　)。
　　A.＜jsp:taglib prefix="s" uri="/struts-tags" %＞

B. <jsp:taglib prefix="s" name="/struts-tags" %>

C. <%@taglib prefix="s" uri="/struts-tags" %>

D. <%@taglib prefix="s" name="/struts-tags" %>

3. 使用(　　)标签进行分支控制。

  A. set          B. param

  C. property        D. if else if else

4. 在 JSP 页面中使用(　　)标签能够实现输出参数值。

  A. <s:set>  B. <s:param>  C. <s:property>  D. <s:text>

5. 使用(　　)标签对集合进行迭代。

  A. <s:out>        B. <s:param>

  C. <s:property>       D. <s:iterator>

6. 简述 Struts 2 标签的分类。

7. 举例说明 Struts 2 中常见的数据标签以及使用方式。

# Struts 2 的输入校验

**主要内容**

（1）手动编程校验。
（2）校验框架校验。

所有用户的输入都是邪恶的，为了保证数据的合法性，输入校验是所有 Web 应用必须处理的问题。Struts 2 框架提供了非常丰富强大的输入校验体系，通过 Struts 2 内置的输入校验器，即可完成大部分的输入校验。因此，通过本章的学习，能够让 Web 应用更加健壮、安全，这也是学习 Struts 2 必备的技术。

## 7.1 手动编程校验

输入校验分为客户端校验和服务器端校验。

客户端校验可以过滤正常用户的误操作，是第一道防线，一般使用 JavaScript 代码实现。仅有客户端校验还是不够的，攻击者还可以绕过客户端校验直接进行非法输入，这样可能会引起系统的异常，为了确保数据的合法性，防止用户通过非正常手段提交错误信息，必须加上服务器端校验。

服务器端校验对于系统的安全性、完整性、健壮性起到了至关重要的作用。Struts 2 框架是非常强大的，它提供了一套验证框架，通过验证框架能够非常简单和快速地完成输入校验。在服务器端，对于输入校验，Struts 2 提供了两种实现方法：一是采用手工编写代码实现（手工编程），二是基于 xml 配置方式实现（Struts 2 校验框架）。

### 7.1.1 核心知识

手动编程主要是通过在类中编写校验逻辑代码实现，有两种实现方式：一是在 Action 类中重写 validate()方法；二是在 Action 类中重写 validateXxx()方法。

**1. 重写 validate()方法**

validate()方法校验 Action 中所有与 execute()方法签名相同的方法。当某个数据校验失败时，在 validate()方法中应该调用 addFieldError()方法向系统的 fieldErrors 添加校验失败信息。为了使用 addFieldError()方法，Action 类需要继承 ActionSupport。

如果系统的 fieldErrors 包含失败信息，Struts 2 将请求转发到名为 input 的 result。在 input 视图中可以通过<s:fielderror/>标签显示失败信息。

下面通过一个简单示例讲解如何重写 validate()方法进行输入校验。

首先，创建一个 JSP 页面 login.jsp，要求验证所有输入项不能为空、密码长度 6~12 位以及两次密码一样等。具体代码如下：

```jsp
<%@page language="java" import="java.util.*" pageEncoding="utf-8"%>
<%@taglib prefix="s" uri="/struts-tags"%>
<%
    String path = request.getContextPath();
    String basePath = request.getScheme() + "://" + request.getServerName() + ":" + request.getServerPort() + path + "/";
%>
<html>
  <head>
    <base href="<%=basePath%>">
  </head>
  <body>
    <s:form action="login.action" theme="simple">
      <table>
        <tr>
          <td>登录账号</td>
          <td><s:textfield name="id"/></td>
        </tr>
        <tr>
          <td>密码</td>
          <td><s:password name="pwd"/></td>
        </tr>
        <tr>
          <td>确认密码</td>
          <td><s:password name="repwd"/></td>
        </tr>
        <tr>
          <td><s:submit value="提交"/></td>
          <td><s:reset value="重置"/></td>
        </tr>
      </table>
      <!-- 显示 this.addFieldError("id", "id不能为空")的信息 -->
      <s:fielderror fieldName="id"/>
      <!-- 显示所有校验失败信息 -->
      <s:fielderror/>
    </s:form>
  </body>
</html>
```

其次，创建 Action 类 LoginAction.java，在该类中重写 validate()方法，具体代码如下：

```java
package action;
import com.opensymphony.xwork2.ActionSupport;
public class LoginAction extends ActionSupport{
    private String id;
```

```java
    private String pwd;
    private String repwd;
    public String getId() {
        return id;
    }
    public void setId(String id) {
        this.id = id;
    }
    public String getPwd() {
        return pwd;
    }
    public void setPwd(String pwd) {
        this.pwd = pwd;
    }
    public String getRepwd() {
        return repwd;
    }
    public void setRepwd(String repwd) {
        this.repwd = repwd;
    }
    /**
     * 实现登录业务处理
     */
    public String execute(){
        return SUCCESS;
    }
    /**
     * 重写validate()方法,进行输入校验,该方法在execute()方法之前执行
     */
    public void validate() {
        if(id == null || id.trim().equals("") ){
            this.addFieldError("id", "id不能为空");
        }
        if(pwd == null || pwd.trim().equals("") ){
            this.addFieldError("pwd", "密码不能为空");
        }else{
            if(pwd.length() < 6 || pwd.length() > 12){
                this.addFieldError("pwdlenth", "密码的长度在 6~12 位");
            }
        }
        if(!pwd.equals(repwd)){
            this.addFieldError("pwdsame", "两次密码不一致");
        }
    }
}
```

再次,配置 Action,具体代码如下:

```xml
<action name = "login" class = "action.LoginAction">
    <result>/index.jsp</result>
    <!-- name = "input"输入校验失败后跳转到该视图,并显示校验失败信息 -->
```

```
        <result name = "input">/login.jsp</result>
    </action>
```

最后，运行 login.jsp 页面，如果输入非法信息，则显示如图 7.1 所示的页面。

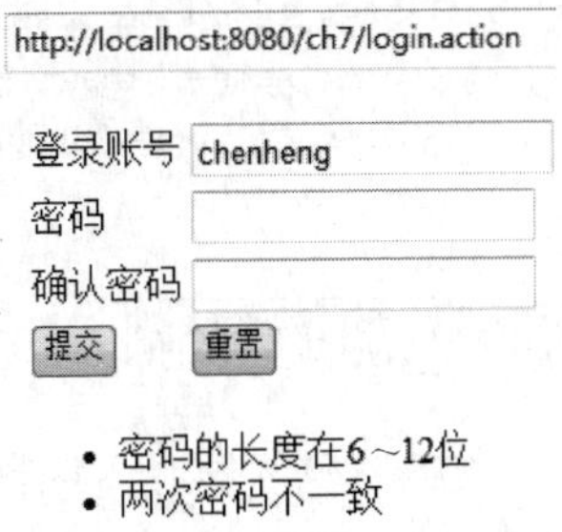

**2. 重写 validateXxx()方法**

validateXxx()只校验 Action 中方法名为 Xxx 的方法，其中 Xxx 的第一个字母要大写。重写 validateXxx()方法进行输入校验与重写 validate()方法基本一样，唯一不同的就是校验的方法名不同而已。

如果将"1. 重写 validate()方法"的示例改成重写 validateXxx()方法进行输入校验，则需要修改 Action 类和配置文件。

图 7.1 重写 validate()方法进行输入校验

修改后的 Action 类的代码如下：

```
package action;
import com.opensymphony.xwork2.ActionSupport;
public class LoginAction extends ActionSupport{
    private String id;
    private String pwd;
    private String repwd;
    //省略 getter 和 setter 方法
    /**
     * 实现登录业务处理
     */
    public String login(){
        return SUCCESS;
    }
    /**
     * 重写 validateLogin()方法,进行输入校验,该方法在 login()方法之前执行
     */
    public void validateLogin() {
        if(id == null || id.trim().equals("")){
            this.addFieldError("id", "id 不能为空");
        }
        if(pwd == null || pwd.trim().equals("")){
            this.addFieldError("pwd", "密码不能为空");
        }else{
            if(pwd.length() < 6 || pwd.length() > 12){
                this.addFieldError("pwdlenth", "密码的长度在 6～12 位");
            }
        }
        if(!pwd.equals(repwd)){
            this.addFieldError("pwdsame", "两次密码不一致");
        }
    }
}
```

配置文件修改后的代码如下：

```
<action name="login" class="action.LoginAction" method="login">
    <result>/index.jsp</result>
    <!-- name="input"输入校验失败后跳转到该视图,并显示校验失败信息 -->
    <result name="input">/login.jsp</result>
</action>
```

### 7.1.2 能力目标

掌握手动编程输入校验的两种实现方式。

### 7.1.3 任务驱动

**1. 任务的主要内容**

编写 JSP 页面 task_7_1.jsp，通过重写 validate()方法对该 JSP 页面的输入项进行校验，要求如下。

① 所有输入项不能为空。
② 年龄必须在 18～65 岁。
③ E-mail 满足正常的格式。
④ JSP 页面的运行效果如图 7.2 所示。

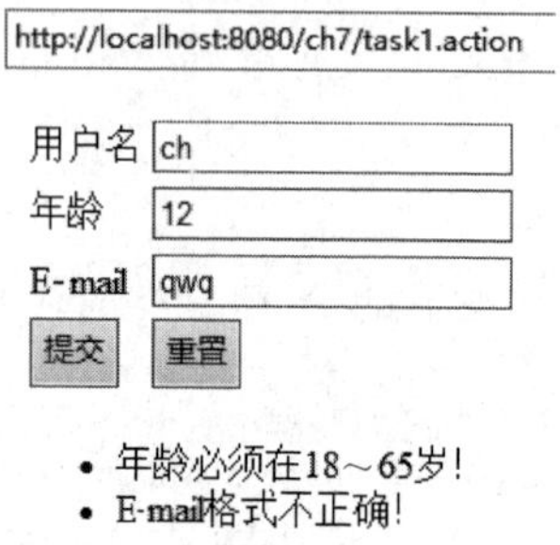

图 7.2  task_7_1.jsp 的校验效果

**2. 任务的代码模板**

task_7_1.jsp 的代码模板如下：

```jsp
<%@ page language="java" import="java.util.*" pageEncoding="utf-8"%>
<%@ taglib prefix="s" uri="/struts-tags" %>
<%
    String path = request.getContextPath();
    String basePath = request.getScheme()+"://"+request.getServerName()+":"+request.getServerPort()+path+"/";
%>
<html>
  <head>
    <base href="<%=basePath%>">
    <title>My JSP 'task_7_1.jsp' starting page</title>
  </head>
  <body>
    <s:form action="task1.action" theme="simple">
      <table>
        <tr>
          <td>用户名</td>
          <td><s:textfield name="uname"/></td>
        </tr>
        <tr>
          <td>年龄</td>
          <td><s:textfield name="uage"/></td>
        </tr>
        <tr>
```

```
                <td>E-mail</td>
                <td><s:textfield name="uemail"/></td>
            </tr>
            <tr>
                <td><s:submit value="提交"/></td>
                <td><s:reset value="重置"/></td>
            </tr>
        </table>
        <!-- 代码1 显示所有校验失败信息 -->
        【代码1】
    </s:form>
  </body>
</html>
```

Task1Action.java 的代码模板如下：

```
package action;
import java.util.regex.Matcher;
import java.util.regex.Pattern;
import com.opensymphony.xwork2.ActionSupport;
public class Task1Action extends ActionSupport{
    private static final long serialVersionUID = 1L;
    private String uname;
    private int uage;
    private String uemail;
    public String execute(){
        return SUCCESS;
    }
    @Override
    public void 【代码2】{
        if(uname.trim().equals("") || uname == null){
            this.addFieldError("uname","用户名不能为空");
        }
        if(uage == 0){
            this.addFieldError("uage","年龄不能为空!");
        }else{
            if(uage<18 || uage>65){
                this.addFieldError("uageScope","年龄必须在18~65岁!");
            }
        }
        if(uemail.trim().equals("") || uemail == null){
            this.addFieldError("uemail","email不能为空!");
        }else{
            //使用正确表达式验证邮箱格式
            String pattern1 = "^([a-z0-9A-Z]+[-|_|\\.]?)+[a-z0-9A-Z]@([a-z0-9A-Z]+(-[a-z0-9A-Z]+)?\\.)+[a-zA-Z]{2,}$";
            Pattern pattern = Pattern.compile(pattern1);
            Matcher mat = pattern.matcher(uemail);
            if (!mat.find()) {
                this.addFieldError("uemailFormat","email格式不正确!");
            }
```

```
        }
    }
    public String getUname() {
        return uname;
    }
    public void setUname(String uname) {
        this.uname = uname;
    }
    public int getUage() {
        return uage;
    }
    public void setUage(int uage) {
        this.uage = uage;
    }
    public String getUemail() {
        return uemail;
    }
    public void setUemail(String uemail) {
        this.uemail = uemail;
    }
}
```

配置文件的代码模板如下：

```
<action name="task1" class="action.Task1Action">
    <result>/index.jsp</result>
    <result【代码3】>/task_7_1.jsp</result>
</action>
```

### 3. 任务小结或知识扩展

通过前面的学习，可以发现 Struts 2 的输入校验需要经过如下几个步骤。

① 类型转换器对请求参数执行类型转换，并把转换后的值赋给 Action 中的属性。

② 如果在执行类型转换的过程中出现异常，系统会将异常信息保存到 ActionContext，conversionError 拦截器将异常信息添加到 fieldErrors 里。不管类型转换是否出现异常，都会进入第③步。

③ 系统通过反射技术先调用 Action 中的 validateXxx() 方法，Xxx 为方法名。

④ 再调用 Action 中的 validate() 方法。

⑤ 经过上述 4 个步骤，如果系统中的 fieldErrors 存在错误信息（即存放错误信息的集合的 size 大于 0），系统自动将请求转发至名称为 input 的视图。如果系统中的 fieldErrors 没有任何错误信息，系统将执行 Action 中的处理方法。

### 4. 任务代码模板的参考答案

【代码 1】<s:fielderror/>

【代码 2】validate()

【代码 3】name="input"

## 7.1.4 实践环节

将 7.1.3 任务中的校验方式修改为"重写 validateXxx()"的校验方式。

## 7.2 校验框架校验

使用 Struts 2 校验框架的好处是将校验逻辑放到配置文件中,实现校验逻辑代码与业务逻辑代码的分离。使用基于校验框架的校验方式实现输入校验时,Action 也需要继承 ActionSupport,并且提供校验文件。同样框架校验的方式也有两种:一是校验 Action 中所有与 execute()方法签名相同的方法;二是校验 Action 中某个方法。

### 7.2.1 核心知识

**1. Struts 2 内置的校验器**

Struts 2 提供了许多默认的校验器。在开发中使用内置校验器能满足大部分的校验需求。这些校验器的定义可以在 xwork-2.x.jar 中的 com.opensymphony.xwork2.validator.validators 下的 default.xml 中找到。Struts 2 框架提供的内置校验器如下。

(1) required:必填校验器,要求 field 的值不能为 null。

(2) requiredstring:必填字符串校验器,要求 field 的值不能为 null,并且长度大于 0,默认情况下会对字符串去前后空格。

(3) stringlength:字符串长度校验器,要求 field 的值必须在指定的范围内,否则校验失败,minLength 参数指定最小长度,maxLength 参数指定最大长度,trim 参数指定校验 field 之前是否去除字符串前后的空格。

(4) regex:正则表达式校验器,检查被校验的 field 是否匹配一个正则表达式。expression 参数指定正则表达式,caseSensitive 参数指定进行正则表达式匹配时,是否区分大小写,默认值为 true。

(5) int:整数校验器,要求 field 的整数值必须在指定范围内,min 指定最小值,max 指定最大值。

(6) double:双精度浮点数校验器,要求 field 的双精度浮点数必须在指定范围内,min 指定最小值,max 指定最大值。

(7) fieldexpression:字段 OGNL 表达式校验器,要求 field 满足一个 OGNL 表达式,expression 参数指定 OGNL 表达式,该逻辑表达式基于 ValueStack 进行求值,返回 true 时校验通过,否则不通过。该校验器只能用于< field-validator >。

(8) email:邮件地址校验器,要求如果 field 的值非空,则必须是合法的邮件地址。

(9) url:网址校验器,要求如果 field 的值非空,则必须是合法的 URL 地址。

(10) date:日期校验器,要求 field 的日期值必须在指定范围内,min 指定最小值,max 指定最大值。

(11) conversion:转换校验器,指定在类型转换失败时,提示的错误信息。

(12) visitor:用于校验 Action 中的复合属性,它指定一个校验文件用于校验复合属性中的属性。

(13) expression:OGNL 表达式校验器,expression 参数指定 OGNL 表达式,该逻辑表达式基于 ValueStack 进行求值,返回 true 时校验通过,否则不通过,该校验器不可用在字段校验器的配置中,只能用于< validator >。

### 2. 常用内置校验器的配置

对于 Struts 2 校验框架来说，一般有两种方式来配置校验器。

（1）使用&lt;validator&gt;。

（2）使用&lt;field-validator&gt;。

当&lt;validator&gt;的子节点中配置了&lt;param name="fieldName"&gt;用于指定对某个属性进行校验时，则达到的效果与&lt;field-validator&gt;是一样的，例如校验 user.id 属性时想使用&lt;validator&gt;来配置，那么写法如下：

```
<validator type="required">
    <param name="fieldName">user.id</param>
    <message>用户 ID 不能为空!</message>
</validator>
```

如果使用&lt;field-validator&gt;配置校验 user.id 属性时，写法如下：

```
<field name="user.id">
    <field-validator type="required">
        <message>用户 ID 不能为空!</message>
    </field-validator>
</field>
```

下面简单介绍一下常用内置校验器的配置示例。

1) required（必填校验器）

```
<field-validator type="required">
    <message>性别不能为空!</message>
</field-validator>
```

2) requiredstring（必填字符串校验器）

```
<field-validator type="requiredstring">
    <param name="trim">true</param>
    <message>用户名不能为空!</message>
</field-validator>
```

3) stringlength（字符串长度校验器）

```
<field-validator type="stringlength">
    <param name="maxLength">12</param>
    <param name="minLength">6</param>
    <message>密码必须在 6～12 位</message>
</field-validator>
```

4) email（邮件地址校验器）

```
<field-validator type="email">
    <message>邮箱格式不正确</message>
</field-validator>
```

5）regex（正则表达式校验器）

```xml
<field-validator type="regex">
    <param name="expression"><![CDATA[^1[3578]\d{9}$ ]]></param>
    <message>手机号格式不正确!</message>
</field-validator>
```

6）int（整数校验器）

```xml
<field-validator type="int">
    <param name="max">100</param>
    <param name="min">0</param>
    <message>年龄必须为 0～100</message>
</field-validator>
```

7）fieldexpression（字段 OGNL 表达式校验器）

```xml
<field-validator type="fieldexpression">
    <param name="expression"><![CDATA[imagefile.length() <= 0]]></param>
    <message>文件不能为空</message>
</field-validator>
```

### 3. 校验 Action 中所有与 execute()签名相同的方法

基于校验框架方式实现对 Action 的所有与 execute()签名相同的方法进行输入校验，校验文件和 Action 类放在同一个包下，文件的取名格式为 ActionClassName-validation.xml，其中 ActionClassName 为 Action 的简单类名，-validation 为固定写法。

下面使用该校验方式对 7.1.3 任务驱动的 JSP 页面 task_7_1.jsp 的输入项进行校验，校验要求与 7.1.3 节一样，具体步骤如下。

（1）创建 JSP 页面 task711721.jsp，该页面与 task-7-1.jsp 页面完全一样，唯一不同的就是表单的 action 属性值。该页面的 action 属性值为 task711721.action。

（2）创建 Action 类 Task711721Action.java，具体代码如下：

```java
package action;
import com.opensymphony.xwork2.ActionSupport;
public class Task711721Action extends ActionSupport{
    private String uname;
    private int uage;
    private String uemail;
    //省略了 setter 和 getter 方法
    public String execute(){
        return SUCCESS;
    }
}
```

（3）在 action 包下编写校验文件 Task711721Action-validation.xml，具体代码如下：

```xml
<?xml version="1.0" encoding="UTF-8"?>
<!DOCTYPE validators PUBLIC "-//OpenSymphony Group//XWork Validator 1.0.3//EN" "http://www.opensymphony.com/xwork/xwork-validator-1.0.3.dtd">
<validators>
```

```xml
<!-- 校验用户名 <field>指定Action中要校验的属性 -->
<field name="uname">
<!-- <field-validator>指定校验器 -->
    <field-validator type="required">
      <!-- <message>为校验失败后的提示信息 -->
        <message>用户名不能为空!</message>
    </field-validator>
</field>
<!-- 校验年龄 -->
<field name="uage">
<!-- 为uage指定两个校验器 -->
    <field-validator type="required">
        <message>年龄不能为空!</message>
    </field-validator>
    <!-- 校验uage的取值范围 -->
    <field-validator type="int">
      <param name="min">18</param>
      <param name="max">65</param>
      <!-- 使用EL表达式 -->
        <message>年龄必须在${min}至${max}之间,当前值为${uage}</message>
    </field-validator>
</field>
<!-- 校验E-mail -->
<field name="uemail">
<!-- 为uemail指定两个校验器 -->
    <field-validator type="required">
        <message>邮箱不能为空!</message>
    </field-validator>
    <!-- 校验uemail的格式 -->
    <field-validator type="email">
        <message>邮箱格式不正确!</message>
    </field-validator>
</field>
</validators>
```

(4) 配置Action,具体代码如下:

```xml
<action name="task711721" class="action.Task711721Action">
    <result>/index.jsp</result>
    <result name="input">/task711721.jsp</result>
</action>
```

(5) 校验显示效果如图7.3所示。

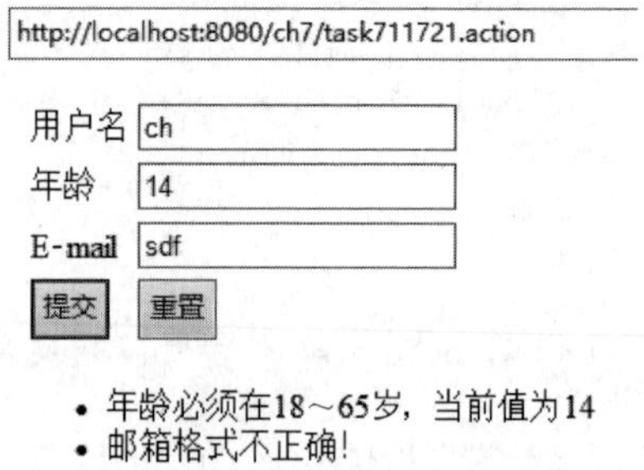

### 4. 校验Action中某个方法

如果只需要对Action中的某个方法实施校验,那么校验文件的取名应为ActionClassName-ActionName-validation.xml,其中ActionName为struts.xml中Action的名称。该校验方式与"3.校验Action中所有与execute()签名相同的方法"校验方式唯一不同的就是校验文件的命名不同,校验文件也是和Action类在同一个目录中。

图7.3 使用校验框架校验Action中所有与execute签名()相同的方法

例如：ManageGoodsAction 中有 add() 和 update() 方法，代码片段如下：

```
…
public String update(){
    return "SUCCESS";
}
public String add(){
    return "SUCCESS";
}
…
```

struts.xml 中的配置如下：

```
<action name="addGoods" class="action.ManageGoodsAction" method="add">
    <result>/index.jsp</result>
    <result name="input">/addGoods.jsp</result>
</action>
<action name="updateGoods" class="action.ManageGoodsAction" method="update">
    <result>/index.jsp</result>
    <result name="input">/updateGoods.jsp</result>
</action>
```

需要对 add() 方法实施验证时，校验文件的取名为：

ManageGoodsAction-addGoods-validation.xml

需要对 update() 方法实施验证时，校验文件的取名为：

ManageGoodsAction-updateGoods-validation.xml

### 7.2.2 能力目标

掌握校验框架输入校验的两种实现方式。

### 7.2.3 任务驱动

**1. 任务的主要内容**

编写 JSP 页面 task_7_2.jsp，通过"基于校验框架方式实现对 Action 的所有与 execute 签名相同的方法校验"对该 JSP 页面的输入项进行校验，要求如下。

① 用户名不能为空，并且长度在 3~10 个字符范围内。

② 使用正则表达式验证手机号。

③ 生日满足日期格式，并且是在 1990-01-01—2017-07-31。

④ JSP 页面的运行效果如图 7.4 所示。

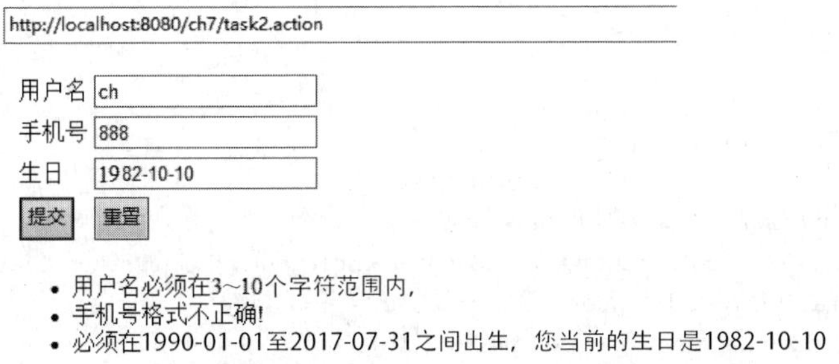

图 7.4　task_7_2.jsp 的校验效果

## 2. 任务的代码模板

task_7_2.jsp 的代码模板如下：

```jsp
<%@ page language="java" import="java.util.*" pageEncoding="utf-8"%>
<%@taglib prefix="s" uri="/struts-tags" %>
<%
    String path = request.getContextPath();
    String basePath = request.getScheme()+"://"+request.getServerName()+":"+request.getServerPort()+path+"/";
%>
<html>
  <head>
    <base href="<%=basePath%>">
    <title>My JSP 'task_7_2.jsp' starting page</title>
  </head>
  <body>
     <s:form action="task2.action" theme="simple">
        <table>
           <tr>
              <td>用户名</td>
              <td><s:textfield name="uname"/></td>
           </tr>
           <tr>
              <td>手机号</td>
              <td><s:textfield name="utel"/></td>
           </tr>
           <tr>
              <td>生日</td>
              <td><s:textfield name="ubirth"/></td>
           </tr>
           <tr>
              <td><s:submit value="提交"/></td>
              <td><s:reset value="重置"/></td>
           </tr>
        </table>
        <!-- 显示所有校验失败信息 -->
        【代码1】
     </s:form>
  </body>
</html>
```

Task2Action.java 的代码如下：

```java
package action;
import model.Task2Model;
import com.opensymphony.xwork2.ActionSupport;
import com.opensymphony.xwork2.ModelDriven;
public class Task2Action extends ActionSupport implements ModelDriven<Task2Model>{
    private static final long serialVersionUID = 1L;
    private Task2Model tsk2 = new Task2Model();
    @Override
```

```java
    public Task2Model getModel() {
        // TODO Auto-generated method stub
        return tsk2;
    }
    public String execute(){
        return SUCCESS;
    }
}
```

Task2Model.java 的代码如下：

```java
package model;
import java.util.Date;
public class Task2Model {
    private String uname;
    private String utel;
    private Date ubirth;
    public String getUname() {
        return uname;
    }
    public void setUname(String uname) {
        this.uname = uname;
    }
    public String getUtel() {
        return utel;
    }
    public void setUtel(String utel) {
        this.utel = utel;
    }
    public Date getUbirth() {
        return ubirth;
    }
    public void setUbirth(Date ubirth) {
        this.ubirth = ubirth;
    }
}
```

校验文件 Task2Action-validation.xml 的代码如下：

```xml
<?xml version="1.0" encoding="UTF-8"?>
<!DOCTYPE validators PUBLIC "-//OpenSymphony Group//XWork Validator 1.0.3//EN" "http://www.opensymphony.com/xwork/xwork-validator-1.0.3.dtd">
<validators>
    <!-- 校验用户名 <field>指定 Action 中要校验的属性 -->
    <field name="uname">
    <!-- <field-validator>指定校验器 -->
        <field-validator type="required">
        <!-- <message>为校验失败后的提示信息 -->
            <message>用户名不能为空!</message>
        </field-validator>
        <field-validator type="stringlength">
            <param name="maxLength">10</param>
```

```xml
            <param name="minLength">3</param>
            <message>用户名必须在3~10个字符范围之内。</message>
        </field-validator>
    </field>
    <!-- 校验手机号 -->
    <field name="utel">
        <!-- 使用正则表达校验utel的格式 -->
        <field-validator type="regex">
            <param name="expression"><![CDATA[^1[3578]\d{9}$]]></param>
            <message>手机号格式不正确!</message>
        </field-validator>
    </field>
    <!-- 校验ubirth -->
    <field name="ubirth">
        <!-- 校验ubirth的格式 -->
        <field-validator type="date">
            <param name="max">2017-07-31</param>
            <param name="min">1990-01-01</param>
            <message>必须在${min}至${max}之间出生,您当前的生日是${ubirth}!</message>
        </field-validator>
    </field>
</validators>
```

配置文件的代码模板如下:

```xml
<action name="task2" class="action.Task2Action">
    <result>/index.jsp</result>
    <result【代码2】>/task_7_2.jsp</result>
</action>
```

### 3. 任务小结或知识扩展

当为某个 Action 既提供了 ActionClassName-validation.xml 校验文件,又提供了 ActionClassName-ActionName-validation.xml 校验文件时,系统将按照下面顺序寻找校验文件:

```
AconClassName-validation.xml
ActionClassName-ActionName-validation.xml
```

系统寻找到第一个校验文件时还会继续搜索后面的校验文件,当搜索到所有校验文件时,会将校验文件里的所有校验规则汇总,然后全部应用于 Action 方法的校验。如果两个校验文件中指定的校验规则冲突,则只使用后面文件中的校验规则。

当某个 Action 继承了另一个 Action,父类 Action 的校验文件将会先被搜索到。假设 ManageGoodsAction 继承 BaseAction。

```xml
<action name="val_*" class="action.ManageGoodsAction" method="{1}"></action>
```

访问上面 Action(val_add)时,系统将会先搜索父类 BaseAction 的校验文件:

```
BaseAction-validation.xml
BaseAction-val_add-validation.xml
```

接着搜索子类 ManageGoodsAction 的校验文件：

ManageGoodsAction-validation.xml
ManageGoodsAction-val_add-validation.xml

应用于上面 Action(val_add)的校验规则为这 4 个文件的总和。

#### 4. 任务代码模板的参考答案

【代码 1】`<s:fielderror/>`

【代码 2】`name="input"`

### 7.2.4 实践环节

将 7.2.3 节任务中的校验方式修改为"基于校验框架方式实现对 Action 的某个方法校验"的方式。

## 7.3 本章小结

本章重点讲解了 Struts 2 的输入校验，包括手动编程方式和校验框架方式。不管哪种校验方式，都需要注意命名规则和校验流程。

## 习 题 7

1. 下列（ ）可实现 Struts 2 的输入校验功能。

    A. 普通 Action 类

    B. 继承自 Action 接口的 Action 类

    C. 继承自 ActionSupport 类的 Action 类

    D. 继承自 ActionValidate 类的 Action 类

2. 简述 Struts 2 中输入校验的流程。

3. 简述 Struts 2 中输入校验的几种方式。

4. 使用校验框架实现输入校验有哪两种方式？它们的命名规则是什么？如果同时使用了两种方式，寻找校验文件的顺序是怎样的？

# Struts 2 的国际化

### 主要内容

(1) Java 国际化的思想。
(2) 资源文件的加载方式。
(3) Struts 2 的国际化。
(4) 用户自定义切换语言。

国际化是商业软件系统的一个基本要求,因为当今的软件系统需要面对全球的浏览者。国际化的目的,就是根据用户的语言环境不同,输出与之相应的页面给用户,以示友好。

Struts 2 的国际化主要有 JSP 页面国际化、校验信息国际化以及 Action 信息国际化等组成。本章主要介绍如何在 Action 中取得国际化消息,以及如何在 JSP 页面中输出国际化消息。最后,本章将示范一个让用户自行选择语言的示例。

## 8.1 程序国际化概述

程序国际化已成为 Web 应用的基本要求。随着网络的发展,大部分的 Web 站点面对的已经不再是本地或者本国的浏览者,而是来自全世界各国各地区的浏览者,因此国际化成为 Web 应用不可或缺的一部分。

#### 1. Java 国际化的思想

Java 国际化的思想是将程序中的信息放在资源文件中,程序根据支持的国家及语言环境,读取相应的资源文件。资源文件是 key-value 对,每个资源文件中的 key 是不变的,但 value 则随不同国家/语言变化。

Java 程序的国际化主要通过两个类来完成。

(1) java.util.Locale:用于提供本地信息,通常称它为语言环境。不同的语言,不同的国家和地区采用不同的 Locale 对象来表示。

(2) java.util.ResourceBundle:该类称为资源包,包含了特定于语言环境的资源对象。当程序需要一个特定于语言环境的资源时(如字符串资源),程序可以从适合当前用户语言环境的资源包中加载它。采用这种方式,可以编写独立于用户语言环境的程序代码,而与特定语言环境相关的信息则通过资源包来提供。

为了实现 Java 程序的国际化,必须事先提供程序所需要的资源文件。资源文件的内容是由很多 key-value 对组成,其中 key 是程序使用的部分,而 value 则是程序界面的显示。

资源文件的命名可以有如下 3 种形式:

(1) baseName.properties。

(2) baseName_language.properties。

(3) baseName_language_country.properties。

baseName 是资源文件的基本名称,由用户自由定义,但是 language 和 country 就必须为 Java 所支持的语言和国家/地区代码。例如:

中国内地　baseName_zh_CN.properties

美国　baseName_en_US.properties

Java 中的资源文件只支持 ISO-8859-1 编码格式字符,直接编写中文会出现乱码。可以使用 Java 命令 native2ascii.exe 解决资源文件的中文乱码问题。使用 MyEclipse 编写资源属性文件,在保存资源文件时,MyEclipse 自动执行 native2ascii.exe 命令。因此,在 MyEclipse 中资源文件不会出现中文乱码问题。

### 2. Java 支持的语言和国家

java.util.Locale 类的常用构造方法如下:

(1) public Locale(String language)。

(2) public Locale(Stringlanguage,String country)。

其中 language 表示语言,它的取值是由小写的两个字母组成的语言代码。country 表示国家或地区,它的取值由大写的两个字母组成的国家或地区代码。

实际上,Java 并不能支持所有国家和语言,如果需要获取 Java 所支持的语言和国家,开发者可以通过调用 Locale 类的 getAvailableLocales()方法获取,该方法返回一个 Locale 数组,该数组里包含了 Java 所支持的语言和国家。

下面的 Java 程序简单示范了如何获取 Java 所支持的国家和语言:

```java
import java.util.Locale;
public class Test {
    public static void main(String[] args) {
        // 返回 Java 所支持的语言和国家的数组
        Locale locales[] = Locale.getAvailableLocales();
        // 遍历数组元素,依次获取所支持的国家和语言
        for (int i = 0; i < locales.length; i++) {
            // 打印出所支持的国家和语言
            System.out.println(locales[i].getDisplayCountry() + " = "
                    + locales[i].getCountry() + " "
                    + locales[i].getDisplayLanguage() + " = "
                    + locales[i].getLanguage());
        }
    }
}
```

程序运行结果如图 8.1 所示。

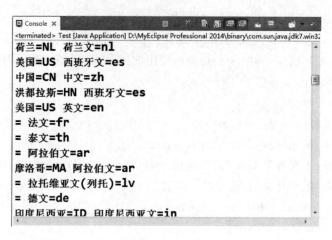

图 8.1　Java 所支持的国家和语言

### 3. Java 程序国际化

假设有如下的简单 Java 程序：

```
public class TestI18N {
    public static void main(String[] args) {
        System.out.println("我要向不同国家的人民问好：您好！");
    }
}
```

为了让该程序支持国际化，需要将"我要向不同国家的人民问好：您好！"对应不同语言环境的字符串，定义在不同的资源文件中。

下面详细介绍在 MyEclipse 下如何创建资源文件，具体步骤如下。

首先，在 Web 应用的 src 目录下新建文件 messageResource_zh_CN.properties 和 messageResource_en_US.properties，创建后效果如图 8.2 所示。

其次，给资源文件添加内容。打开资源文件，切换到 Properties 选项卡，单击 Add 按钮打开添加 Add Property 窗口，在该窗口的 Name * 文本框输入 key（程序使用的部分），在该窗口的 Value 文本框输入 value（要国际化的信息），如图 8.3 所示。

图 8.2　创建资源文件

图 8.3　添加资源文件内容

最后，单击图 8.3 中的 Finish 按钮，并保存 messageResource_zh_CN.properties 文件，然后切换到 Source 选项卡，可看到如图 8.4 所示的效果。

```
hello=\u6211\u8981\u5411\u4E0D\u540C\u56
```

图 8.4　Unicode 编码资源文件

图 8.4 显示的内容看似是很多乱码，实际是 Unicode 编码文件内容。至此，资源文件 messageResource_zh_CN.properties 创建完成。同理，创建资源文件 messageResource_en_US.properties，具体步骤不再赘述。

现在将 TestI18N.java 程序修改成如下形式：

```java
import java.util.Locale;
import java.util.ResourceBundle;
public class TestI18N {
    public static void main(String[] args) {
        //取得系统默认的国家语言环境
        Locale lc = Locale.getDefault();
        //根据国家语言环境加载资源文件
        ResourceBundle rb = ResourceBundle.getBundle("messageResource", lc);
        //打印出从资源文件中取得的信息
        System.out.println(rb.getString("hello"));
    }
}
```

上面程序中的打印语句打印的内容是从资源文件中读取的信息。如果在中文环境下运行程序，将打印"我要向不同国家的人民问好：您好！"；如果在"控制面板"中将机器的语言环境设置成美国，然后再次运行该程序，将打印"I want to say hello to all world！"。

需要注意的是，如果程序找不到对应国家/语言的资源文件时，系统该怎么办？假设以简体中文环境为例，先搜索如下文件：

messageResource_zh_CN.properties

如果没有找到国家/语言都匹配的资源文件，再搜索语言匹配文件，即搜索如下文件：

messageResource_zh.properties

如果上面的文件还没有搜索到，则搜索 baseName 匹配的文件，即搜索如下文件：

messageResource.properties

如果上面 3 个文件都找不到，则系统将出现异常。

### 8.1.2　能力目标

理解程序国际化的思想，掌握 Java 程序国际化的实现方式。

### 8.1.3　任务驱动

**1. 任务的主要内容**

编写一个 Java 应用程序 Task1。如果系统是中文环境，则该程序从资源文件

messageResource_zh_CN.properties 中读取消息文本"task1＝中文。"；如果系统是英文环境，则从英文资源文件 messageResource_en_US.properties 中读取消息文本"task1＝English."。

### 2. 任务的代码模板
Task1.java 的代码模板如下：

```java
import java.util.Locale;
import java.util.ResourceBundle;
public class Task1 {
    public static void main(String[] args) {
        //取得系统默认的国家语言环境
        Locale lc = Locale.getDefault();
        //根据国家语言环境加载资源文件
        ResourceBundle rb = 【代码1】
        //打印出从资源文件中取得的信息
        System.out.println(【代码2】);
    }
}
```

### 3. 任务小结或知识扩展
带占位符的国际化信息

在资源文件中的消息文本可以带有参数，例如：

welcome＝{0}，欢迎学习 Struts 2。

花括号中的数字是一个占位符，可以被动态的数据替换。在消息文本中的占位符可以使用 0~9 的数字，也就是说，消息文本的参数最多可以有 10 个。例如：

welcome＝{0}，欢迎学习 Struts 2，今天是星期{1}。

要替换消息文本中的占位符，可以使用 java.text.MessageFormat 类，该类提供了一个静态方法 format()，用来格式化带参数的文本，format()方法定义如下：

```java
public static String format(String pattern, Object ...arguments)
```

其中，pattern 字符串就是一个带占位符的字符串，消息文本中的数字占位符将按照方法参数的顺序（从第二个参数开始）被替换。

替换占位符的示例代码如下：

```java
import java.text.MessageFormat;
import java.util.Locale;
import java.util.ResourceBundle;
public class TestFormat {
    public static void main(String[] args) {
        //取得系统默认的国家语言环境
        Locale lc = Locale.getDefault();
        //根据国家语言环境加载资源文件
        ResourceBundle rb = ResourceBundle.getBundle("messageResource", lc);
        //从资源文件中取得的信息
        String msg = rb.getString("welcome");
```

```
        //替换消息文本中的占位符,消息文本中的数字占位符将按照参数的顺序
        //(从第二个参数开始)被替换,即"我"替换{0},"5"替换{1}
        String msgFor = MessageFormat.format(msg, "我","5");
        System.out.println(msgFor);
    }
}
```

**4. 任务代码模板的参考答案**

【代码 1】`ResourceBundle.getBundle("messageResource", lc);`

【代码 2】`rb.getString("task1")`

### 8.1.4 实践环节

编写一个 Java 应用程序 Pratice814。如果系统是中文环境,则该程序从资源文件 messageResource_zh_CN.properties 中读取消息文本 "practice814＝今天{0}很高兴,{1}也不错,明天就是星期{2}了。";如果系统是英文环境,则从英文资源文件 messageResource_en_US.properties 中读取消息文本 "practice814＝ Today,{0} is very glad,{1} is too good, tomorrow will be {2}。"。

## 8.2　Struts 2 的国际化方法

Struts 2 的国际化是建立在 Java 国际化基础之上,Struts 2 框架的底层国际化与 Java 国际化是一致的,作为一个良好的 MVC 框架,Struts 2 将 Java 国际化的功能进行了封装和简化,开发者使用起来会更加简单快捷。

### 8.2.1 核心知识

**1. Struts 2 中加载全局资源文件**

Struts 2 提供了许多加载资源文件的方式,最方便、最常用的就是加载全局资源文件,加载全局资源文件的方式是通过在 struts.xml 文件中配置常量 struts.custom.i18n.resources 来实现,该常量值为全局资源文件的 baseName。

一旦指定了全局资源文件,即可实现程序的国际化。假设 baseName 为 messageResource,则在 struts.xml 文件中配置如下一个常量:

`<constant name="struts.custom.i18n.resources" value="messageResource"/>`

通过这种方式加载全局资源文件后,Struts 2 应用就可以在所有地方取得这些资源文件了,包括 JSP 页面、Action 和校验文件。

**2. 国际化信息输出**

国际化信息输出主要有如下几种方式。

1) 在 JSP 页面中

为了在 JSP 页面中输出国际化信息,可以使用 Struts 2 的 `<s:text.../>` 标签,该标签可以指定一个 name 属性,该属性指定了国际化资源文件中的 key。

2) 在表单标签中

通过 key 属性指定资源文件中的 key, 如：

<s:textfield name = "realname" key = "username"/>

或者

<s:textfield name = "realname" label = "%{getText('username')}"/>

3) 在 Action 类中

Action 类可以继承 ActionSupport, 使用 getText() 方法取得国际化信息, 该方法的第一个参数用于指定资源文件中的 key。

4) 在校验文件中

在校验文件中, 可以通过 message 的 key 属性指定资源文件中的 key, 如：

<message key = "login.error.username"/>

替换占位符的方式有两种：

(1) 在 JSP 页面中替换占位符。

在 JSP 页面中输出带占位符的国际化信息, 例如, 资源文件内容如下：

welcome={0},欢迎学习 Struts 2,今天是星期{1}。

则在 JSP 页面中可这样输出带占位符的国际化信息：

```
<s:text name = "welcome">
    <!-- <s:text name = "chenheng"/>替换{0} -->
    <s:param><s:text name = "chenheng"/></s:param>
    <!-- <s:text name = "monday"/>替换{1} -->
    <s:param><s:text name = "monday"/></s:param>
</s:text>
```

(2) 在 Action 类中替换占位符。

在 Action 类中获取带占位符的国际化信息, 可以使用 getText(String name, String[] args)或 getText(String name, List args)方法。该方法的第二个参数既可以是一个字符串数组, 也可以是字符串组成的 List 对象, 从而完成对占位符的填充。其中字符串数组、字符串集合中第一个元素将填充第一个占位符, 字符串数组、字符串集合中第二元素将填充第二个占位符, 以此类推。

下面通过一个具体示例讲解如何输出国际化信息, 步骤如下。

① 在资源文件 messageResource_zh_CN.properties 中添加如下国际化信息：

```
uname = 用户名
upass = 密码
login = 登录
mysubmit = 提交
login.no.null = {0}不能为空！
chenheng = 陈恒
welcome = {0},欢迎学习 Struts 2,今天是星期{1}。
monday = 星期一
```

上述国际化信息对应的英文如下:

```
uname = username
upass = userpass
login = login
mysubmit = submit
login.no.null = {0} is not null.
chenheng = chenheng
welcome = {0},welcome to study struts2.Today is {1}.
monday = monday
```

② 创建 JSP 页面 login.jsp,在页面中输出国际化信息,具体代码如下:

```jsp
<%@ page language="java" import="java.util.*" pageEncoding="utf-8"%>
<%@taglib prefix="s" uri="/struts-tags" %>
<html>
  <head>
    <title>My JSP 'login.jsp' starting page</title>
  </head>
  <body>
    <s:form action="login.action" method="post">
    <!-- s:text 输出国际化消息,name 指向资源文件中的 key -->
    <h3><s:text name="login"/></h3>
    <!-- 在表单元素中,key 指向资源文件中的 key -->
    <s:textfield name="username" key="uname"/>
    <s:password name="password" key="upass"/>
    <s:submit value="%{getText('mysubmit')}"/>
    <s:fielderror/>
    </s:form>
  </body>
</html>
```

③ 创建 Action 类 LoginAction.java,具体代码如下:

```java
package action;
import com.opensymphony.xwork2.ActionSupport;
public class LoginAction extends ActionSupport{
    private String username;
    private String password;
    public String getUsername() {
        return username;
    }
    public void setUsername(String username) {
        this.username = username;
    }
    public String getPassword() {
        return password;
    }
    public void setPassword(String password) {
        this.password = password;
    }
    public String execute(){
        return SUCCESS;
    }
    @Override
```

```java
public void validate() {
    if(username == null || username.trim().equals("") ){
        //在Action中使用getText方法取得资源文件的信息,方法参数指向资源文件里
        //的keygetText("uname")替换{0}
        String a[] = {getText("uname")};
        this.addFieldError("usernamenull", getText("login.no.null", a));
    }
    if(password == null || password.trim().equals("") ){
        //在Action中使用getText方法取得资源文件的信息,getText("upass")替换{0}
        String a[] = {getText("upass")};
        this.addFieldError("passwordnull", getText("login.no.null", a));
    }
}
```

④ 配置Action,具体代码如下:

```xml
<action name="login" class="action.LoginAction">
    <result>/index.jsp</result>
    <result name="input">/login.jsp</result>
</action>
```

⑤ 修改index.jsp页面,具体代码如下:

```jsp
<%@ page language="java" import="java.util.*" pageEncoding="utf-8"%>
<%@taglib prefix="s" uri="/struts-tags" %>
<html>
  <head>
  </head>
  <body>
    <s:text name="welcome">
    <!-- <s:text name="chenheng"/>替换{0} -->
    <s:param><s:text name="chenheng"/></s:param>
    <!-- <s:text name="monday"/>替换{1} -->
    <s:param><s:text name="monday"/></s:param>
    </s:text>
  </body>
</html>
```

如果用户将自己的浏览器语言设置成"英国/美国",运行上述示例中的login.jsp页面,单击submit按钮后,将看到如图8.5所示的页面。

图8.5 英国/美国语言环境下的login.jsp页面

在图 8.5 中输入任意的用户名和密码,登录成功,进入 index.jsp,将看到如图 8.6 所示的页面。

图 8.6　英国/美国语言环境下的 index.jsp 页面

**注意**:读者学习 8.3 节后,需要输出不同语言的国际化信息时,不用再修改浏览器的语言设置了。

### 8.2.2　能力目标

理解 Struts 2 国际化的原理,掌握国际化信息的输出方式。

### 8.2.3　任务驱动

#### 1. 任务的主要内容

编写一个 JSP 页面 task_8_2.jsp,在该页面中输出国际化消息"task2=我是{0},我正在学习{1}"。

#### 2. 任务的代码模板

task_8_2.jsp 的代码模板如下:

```jsp
<%@ page language="java" import="java.util.*" pageEncoding="UTF-8"%>
<%@taglib prefix="s" uri="/struts-tags" %>
<%
    String path = request.getContextPath();
    String basePath = request.getScheme() + "://" + request.getServerName() + ":" + request.getServerPort() + path + "/";
%>
<!DOCTYPE HTML PUBLIC "-//W3C//DTD HTML 4.01 Transitional//EN">
<html>
  <head>
    <base href="<%=basePath%>">
    <title>My JSP 'task_8_2.jsp' starting page</title>
  </head>
  <body>
    <!-- 代码 1 取出资源属性文件中 task2 消息 -->
    【代码 1】
        <!-- 代码 2 取出资源属性文件中 chenheng 消息,替换{0} -->
        【代码 2】
        <!-- 代码 3 取出资源属性文件中 math 消息,替换{1} -->
        【代码 3】
    </s:text>
  </body>
</html>
```

### 3. 任务小结或知识扩展

Struts 2 中除了指定加载全局资源文件外，Struts 2 还提供了多种方式加载资源文件，包括指定包范围资源文件、Action 类范围资源文件以及临时指定资源文件等，并且对于不同范围的资源文件有不同的加载顺序。

1) 包范围资源文件

在一个大型应用中，整个应用有大量的内容需要实现国际化，如果将国际化的内容都放置在全局资源属性文件中，显然会导致资源文件变得过于庞大、臃肿，不便于维护，这个时候可以针对不同模块，使用包范围来组织国际化文件，具体步骤如下。

① 首先编写中英文资源文件，命名方式：package_language_country.properties，保存目录为该包的根目录。包范围资源文件的 baseName 就是 package，不是 Action 所在的包名。一旦创建了这个范围的国际化资源文件，应用中处于该包下（包括子包）的所有 Action 都可以访问该资源文件。

② 当查找指定 key 的国际化信息时，系统会先从 package 资源文件查找，当找不到对应的 key 时，沿着当前包上溯，一直找到顶层包，如果还没找到对应的 key 时，才会从常量 struts.custom.i18n.resources 指定的全局资源文件中寻找。

2) Action 类范围资源文件

Struts 2 还允许单独为 Action 类指定一份国际化资源文件。Action 类范围的资源文件只能被该 Action 类访问，具体步骤如下。

① 首先编写中英文资源文件，命名方式：

ActionClassName_language_country.properties

其中 ActionClassName 为 Action 类的简单名称（不含有包名），保存在 Action 同目录下。

② 当查找指定 key 的国际化信息时，系统会先从 ActionClassName_language_country.properties 资源文件查找，如果没有找到对应的 key，然后沿着当前包往上查找基本名为 package(Action 类所在的包)的包范围资源文件，一直找到顶层包。如果还没有找到对应的 key，最后才从常量 struts.custom.i18n.resources 指定的全局资源文件中寻找。

3) 临时指定资源文件

在 JSP 页面中输出国际化信息时，可以借助于 <s:i18n.../> 标签临时指定资源文件的位置。

如果将 <s:i18n.../> 标签作为 <s:text.../> 标签的父标签，则 <s:text.../> 标签将会直接加载 <s:i18n.../> 标签里指定的国际化资源文件；如果将 <s:i18n.../> 标签作为表单标签的父标签，则表单标签的 key 属性将会从国际化资源文件中加载该信息。

假设在 Web 应用 src 目录下新建中英文两份资源文件，第一份是 tmp_zh_CN.properties，该文件的内容是：

```
tmp1 = 临时信息 1
tmp2 = 临时信息 2
```

第二份是 tmp_en_US.properties，该文件的内容是：

```
tmp1 = tmp1
tmp2 = tmp2
```

下面是测试临时资源文件的 JSP 页面 test.jsp,具体代码如下:

```jsp
<%@ page language="java" import="java.util.*" pageEncoding="utf-8"%>
<%@taglib prefix="s" uri="/struts-tags" %>
<html>
  <head>
    <title>My JSP 'test.jsp' starting page</title>
  </head>
  <body>
    <!-- 将 i18n 作为 s:text 标签的父标签,临时指定资源文件的 baseName 为 tmp -->
    <s:i18n name="tmp">
    <!-- 输出国际化信息 -->
    <s:text name="tmp1"/>
    <s:text name="tmp2"/>
    </s:i18n>
    <!-- 将 i18n 作为 s:form 标签的父标签,临时指定资源文件的 baseName 为 tmp -->
    <s:i18n name="tmp">
    <s:form>
        <s:textfield name="uname" key="tmp1"/>
        <s:textfield name="uage" key="tmp2"/>
    </s:form>
    </s:i18n>
  </body>
</html>
```

**4. 任务代码模板的参考答案**

【代码 1】`<s:text name="task2">`

【代码 2】`<s:param><s:text name="chenheng"/></s:param>`

【代码 3】`<s:param><s:text name="math"/></s:param>`

### 8.2.4 实践环节

将任务中 task_8_2.jsp 页面修改为"借助于`<s:i18n.../>`标签临时指定资源文件的位置"方式输出国际化消息。

## 8.3 用户自定义切换语言示例

在许多成熟的商业软件系统中,可以让用户自由切换语言,而不是修改浏览器的语言设置。一旦用户选择了自己需要使用的语言环境,整个系统的语言环境将一直是这种语言环境。Struts 2 也可以允许用户自行选择程序语言。

为了简化设置语言环境,Struts 2 提供了一个名为 i18n 的拦截器,该拦截器注册在默认的拦截器栈中。i18n 拦截器在执行 Action 方法前,自动查找请求中一个名为 request_locale 的参数。如果该参数存在,拦截器就将其作为参数,转换成 Locale 对象,并将其设为用户默认的 Locale(国家/语言环境)。除此之外,i18n 拦截器还会将上面生成的 Locale 对象以"WW_TRANS_I18N_LOCALE"为属性名,保存在用户的 session 对象中。一旦用户 session 中存在一个名为"WW_TRANS_I18N_LOCALE"的属性,则该属性指定的 Locale 将会作为浏览者的默认国家/语言环境。

但需要注意的是,i18n 拦截器只有请求 Action 的时候才执行,因此单纯的页面跳转并没有执行 i18n 拦截器。这样问题就来了,单纯的页面跳转如何国际化? 为了解决该问题,本书采用过滤器的方式满足用户自定义切换语言,具体步骤如下。

(1) 编写 MyRequestWrapper 类,在该类中获得默认的 Locale 对象,具体代码如下:

```java
package filter;
import java.util.Locale;
import javax.servlet.http.HttpServletRequest;
import javax.servlet.http.HttpServletRequestWrapper;
import javax.servlet.http.HttpSession;
public class MyRequestWrapper extends HttpServletRequestWrapper {
    private Locale locale = null;
    public MyRequestWrapper(HttpServletRequest request) {
        super(request);
        HttpSession session = request.getSession();
        //...............start...............
        /* 下面 if 语句解决国际化入口页面不经过 Action 时的情形,
         * 例如 <a href = "selectLanguage.jsp?request_locale = zh_CN">.
         * 如果通过 Action 访问国际化入口页面,可以不需要此处的 if 语句
         */
        String local = request.getParameter("request_locale");
        if(local != null){
            String a[] = local.split("_");
            //将国家/语言环境保存到 session 中
            session.setAttribute("WW_TRANS_I18N_LOCALE",new Locale(a[0],a[1]));
        }
        //...............end...............
        //获得浏览者的默认 Locale(代表国家/语言环境)
        locale = (Locale) session.getAttribute("WW_TRANS_I18N_LOCALE");
    }
    @Override
    public Locale getLocale() {
        if (locale != null) {
            return locale;
        }
        return super.getLocale();
    }
}
```

(2) 编写 I18NFilter 过滤器类,在该类中将 request 对象进行处理,具体代码如下:

```java
package filter;
import java.io.IOException;
import javax.servlet.Filter;
import javax.servlet.FilterChain;
import javax.servlet.FilterConfig;
```

```java
import javax.servlet.ServletException;
import javax.servlet.ServletRequest;
import javax.servlet.ServletResponse;
import javax.servlet.http.HttpServletRequest;
public class I18NFilter implements Filter{
    @Override
    public void destroy() {
    }
    @Override
    public void doFilter(ServletRequest req, ServletResponse resp,
            FilterChain filterChain) throws IOException, ServletException {
        HttpServletRequest r = (HttpServletRequest) req;
        MyRequestWrapper request = new MyRequestWrapper(r);
        filterChain.doFilter(request, resp);
    }
    @Override
    public void init(FilterConfig arg0) throws ServletException {

    }
}
```

（3）在 WEB-INF/web.xml 文件中配置过滤器，具体代码如下：

```xml
<filter>
  <filter-name>I18Nfrilter</filter-name>
  <filter-class>filter.I18NFilter</filter-class>
</filter>
<filter-mapping>
  <filter-name>I18Nfrilter</filter-name>
  <url-pattern>/*</url-pattern>
</filter-mapping>
```

（4）在 Web 应用中进行自定义切换语言测试，测试过程具体如下。

① 在资源文件 messageResource_zh_CN.properties 中添加如下需要国际化的信息：

```
zhcn=简体中文
usen=美式英语
next=下一页
```

其他用到的国际化信息，在前面已经添加，此处不再添加。英文资源文件 messageResource_en_US.properties 中添加如下需要国际化的信息：

```
zhcn=Simplified Chinese
usen=American English
next=next
```

② 编写程序入口页面 selectLanguage.jsp，在该页面中自定义切换语言，具体代码如下：

```jsp
<%@ page language="java" import="java.util.*" pageEncoding="utf-8" %>
<%@ taglib prefix="s" uri="/struts-tags" %>
<html>
```

```
<body>
    <a href = "selectLanguage.action?request_locale = zh_CN"><s:text name = "zhcn"/></a>
    <br><br>
    <!-- ?前为空相当于链接到当前页面,与 selectLanguage.jsp?request_locale = en_US 效果一样 -->
    <a href = "?request_locale = en_US"><s:text name = "usen"/></a>
    <br><br>
    <a href = "input.jsp"><s:text name = "next"/></a>
</body>
</html>
```

③ 编写 Action 类,处理 selectLanguage.action 请求,具体代码如下:

```
package action;
import com.opensymphony.xwork2.ActionSupport;
public class SelectLanguageAction extends ActionSupport{
    public String execute(){
        return SUCCESS;
    }
}
```

④ 配置 Action,具体代码如下:

```
<action name = "selectLanguage" class = "action.SelectLanguageAction">
    <result>/selectLanguage.jsp</result>
</action>
```

⑤ 编写 input.jsp 页面,在页面 selectLanguage.jsp 中单击"下一页"按钮进入该页面,具体代码如下:

```
<%@ page language = "java" import = "java.util.*" pageEncoding = "utf-8"%>
<%@taglib prefix = "s" uri = "/struts-tags" %>
<html>
  <head>
  </head>
  <body>
    <s:text name = "uname"/><br><br>
    <a href = "index.jsp"><s:text name = "next"/></a>
  </body>
</html>
```

⑥ 编写 index.jsp 页面,在页面 input.jsp 中单击"下一页"按钮进入该页面,具体代码如下:

```
<%@ page language = "java" import = "java.util.*" pageEncoding = "utf-8"%>
<%@taglib prefix = "s" uri = "/struts-tags" %>
<html>
  <head>
  </head>
  <body>
    <s:text name = "welcome">
    <!-- <s:text name = "Chenheng"/>替换{0} -->
```

```
          <s:param><s:text name = "Chenheng"/></s:param>
          <!-- <s:text name = "monday"/>替换{1} -->
          <s:param><s:text name = "monday"/></s:param>
        </s:text>
        <br><br>
        <a href = "selectLanguage.jsp"><s:text name = "next"/></a>
      </body>
    </html>
```

测试上面程序时,浏览者不需要对浏览器做任何设置修改。运行程序入口页面 selectLanguage.jsp,并单击各页面中的"下一页"按钮,效果如图 8.7 所示。

图 8.7 自定义切换语言(中文环境)

单击图 8.7 中的"美式英语"按钮,并单击各页面中的"下一页"按钮,效果如图 8.8 所示。

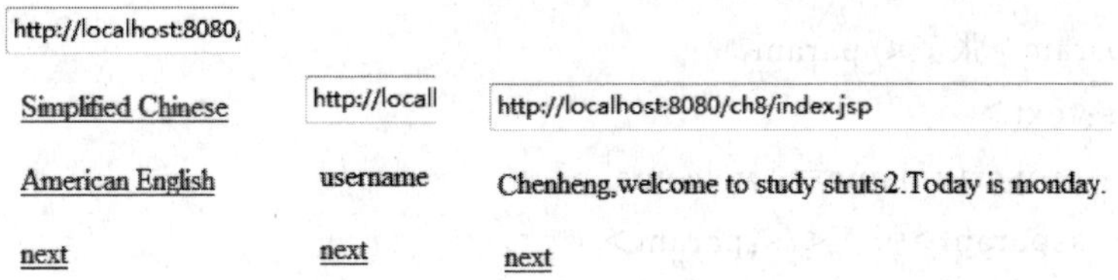

图 8.8 自定义切换语言(英文环境)

## 8.4 本章小结

本章主要讲解了 Struts 2 的国际化知识,Struts 2 的国际化主要由校验提示信息国际化、Action 信息国际化和 JSP 页面信息国际化等组成。本章详细讲述了国际化资源文件的加载方式,包括全局资源文件、包范围资源文件和 Action 类范围资源文件。最后,本章给出了一个让用户自行选择语言的示例,介绍了 Struts 2 国际化的内在原理。

## 习 题 8

1. 在 JSP 页面中可以通过 Struts 2 提供的(   )标签来输出国际化信息。
   A. <s:input>            B. <s:message>
   C. <s:submit>           D. <s:text>
2. 资源文件的后缀名为(   )。
   A. txt          B. doc          C. property          D. properties

3. 国际化资源文件中有 welcome＝"欢迎"，在 JSP 页面中使用哪个标签输出其国际化信息（　　）。

  A．＜s:text name＝"welcome"/＞

  B．＜s:getText name＝"welcome"/＞

  C．＜s:property name＝"welcome"/＞

  D．＜s:text value＝"welcome"/＞

4. 国际化资源文件中有 welcome＝"欢迎"，在 Action 类中可以使用（　　）方法实现国际化。

  A．text("welcome");      B．getText("welcome");

  C．outText("welcome");     D．outI18N("welcome");

5. 在 Struts 表单标签中使用哪个属性实现国际化（　　）。

  A．name   B．value   C．key   D．id

6. 国际化资源文件中有"welcome＝{0}，你好，欢迎到来"，在 JSP 页面中输出带占位符的国际化信息的方法是（　　）。

  A．＜s:text name＝"welcome"＞

    ＜s:param＞张三＜/s:param＞

    ＜/s:text＞

  B．＜s:text name＝"welcome"＞

    ＜param＞张三＜/param＞

    ＜/s:text＞

  C．＜s:property name＝"welcome"＞

    ＜s:param＞张三＜/s:param＞

    ＜/s:property＞

  D．＜s:text value＝"welcome" param＝"张三"/＞

7. 什么是国际化？国际化资源文件的命名格式是怎样的？

8. 在 JSP 页面中使用＜s:text＞标签输出国际化信息；在 Action 类中如何输出国际化信息？表单标签中又当如何？

9. 加载国际化资源文件有哪 4 种方式？

# 文件的上传和下载

**主要内容**

(1) Struts 2 单文件上传。
(2) Struts 2 多文件上传。
(3) Struts 2 文件下载。

文件上传是 Web 应用经常需要面对的问题。对于 Java 应用而言上传文件有多种方式，包括使用文件流手工编程上传、基于 commons-fileupload 组件的文件上传、基于 Servlet 3.0 的文件上传等方式。本章将重点介绍如何使用 Struts 2 框架进行文件上传。

## 9.1 Struts 2 文件上传

Struts 2 默认使用的是基于 commons-fileupload 组件的文件上传，只不过 Struts 2 在原有的文件上传组件上做了进一步封装，简化了文件上传的代码实现，取消了不同上传组件上的编程差异。

### 9.1.1 核心知识

**1. 基于表单的文件上传**

表单标签：<s:file/>，在浏览器中会显示一个文本框和一个按钮，文本框可供用户填写本地文件的文件名和路径名，按钮可以让浏览器打开一个文件选择框供用户选择文件。

文件上传的表单例子如下：

```
<s:form action="upload.action" method="post" enctype="multipart/form-data">
    <s:file name="headImage"/>
    <s:submit>
</s:form>
```

基于表单的文件上传，不要忘记使用 enctype 属性，并将它的值设置为 multipart/form-data。同时，表单的提交方式设置为 post。为什么需要这样呢？下面从 enctype 属性说起。

表单的 enctype 属性指定的是表单数据的编码方式，该属性有如下 3 个值。

（1）application/x-www-form-urlencoded：这是默认的编码方式，它只处理表单域里的 value 属性值。

(2) multipart/form-data：这种编码方式会以二进制流的方式来处理表单数据，这种编码方式会将文件域指定文件的内容也封装到请求参数里。

(3) text/plain：当表单的 action 属性为 mailto：URL 的形式时才使用这种编码方式，该方式主要适用于直接通过表单发送邮件的方式。

由上面 3 个属性的解释可知，基于表单上传文件时，enctype 的属性值应为 multipart/form-data。需要注意的是，一旦设置了表单的 enctype="multipart/form-data" 属性，就将无法通过 HttpServletRequest 对象的 getParameter()方法取得请求参数。

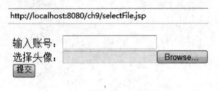

图 9.1 文件上传页面

### 2. Struts 2 单文件上传

本节通过一个示例讲解 Struts 2 如何实现单文件的上传。假设有如图 9.1 所示的文件上传页面，为了完成文件上传，应该将该页面表单的 enctype 属性设置为 multipart/form-data。该页面（selectFile.jsp）的代码如下：

```jsp
<%@ page language="java" import="java.util.*" pageEncoding="utf-8" %>
<%@ taglib prefix="s" uri="/struts-tags" %>
<html>
  <head>
    <title>My JSP 'selectFile.jsp' starting page</title>
  </head>
  <body>
    <s:form action="upload.action" method="post" theme="simple" enctype="multipart/form-data">
      输入账号：<s:textfield name="uid"/><br>
      选择头像：<s:file name="headImage"/><br>
      <s:submit value="提交"/>
    </s:form>
  </body>
</html>
```

当文件上传页面提交请求时，请求发送到 upload.action，这是一个 Struts 2 的 Action，该 Action 处理上传请求，具体的 UploadAction 类代码如下：

```java
package action;
import java.io.File;
import java.io.IOException;
import org.apache.commons.io.FileUtils;
import org.apache.struts2.ServletActionContext;
import com.opensymphony.xwork2.ActionSupport;
public class UploadAction extends ActionSupport{
    private static final long serialVersionUID = 1L;
    //封装账号(uid)请求参数的属性
    private String uid;
    //封装上传文件域的属性
    private File headImage;
    //封装上传文件类型的属性
    private String headImageContentType;
```

```java
//封装上传文件名的属性
private String headImageFileName;
public String execute() throws IOException{
    //上传文件的保存位置"/images",该位置是指tomcat的webapps\ch9\images,发布后使用
    String realpath1 = ServletActionContext.getServletContext().getRealPath("/images");
    //该位置是指"D:/Struts2 do workspace/ch9/WebRoot/upimages",开发时使用
    String realpath2 = "D:/Struts2 do workspace/ch9/WebRoot/upimages";
    //声明文件目录 images
    File file1 = new File(realpath1);
    File file2 = new File(realpath2);
    //如果目录不存在,则创建该目录
    if(!file1.exists()){
        file1.mkdirs();
    }
    //如果目录不存在,则创建该目录
    if(!file2.exists()){
        file2.mkdirs();
    }
    //实现文件上传
    FileUtils.copyFile(headImage, new File(file1, headImageFileName));
    FileUtils.copyFile(headImage, new File(file2, headImageFileName));
    return SUCCESS;
}
public String getUid() {
    return uid;
}
public void setUid(String uid) {
    this.uid = uid;
}
public File getHeadImage() {
    return headImage;
}
public void setHeadImage(File headImage) {
    this.headImage = headImage;
}
public String getHeadImageContentType() {
    return headImageContentType;
}
public void setHeadImageContentType(String headImageContentType) {
    this.headImageContentType = headImageContentType;
}
public String getHeadImageFileName() {
    return headImageFileName;
}
public void setHeadImageFileName(String headImageFileName) {
    this.headImageFileName = headImageFileName;
}
}
```

值得注意的是,上面的Action除了包含两个表单域的name属性外,还包含了两个属性:headImageContentType和headImageFileName,这两个属性分别用于封装上传文件的

文件类型、上传文件的文件名。可以这样认为：如果表单中包含一个 name 属性为×××的文件域，则对应的 Action 需要使用 3 个属性来封装文件域的信息。

（1）类型为 java.io.File 的×××属性封装了该文件域对应的文件内容。
（2）类型为 String 的×××FileName 属性封装了该文件域对应的文件的文件名。
（3）类型为 String 的×××ContentType 属性封装了该文件域对应的文件的文件类型。

所以在 Action 的 execute 方法中，可以直接通过这 3 个属性获取上传文件的文件名、文件类型和文件内容。

配置 UploadAction，具体代码如下：

```xml
<?xml version = "1.0" encoding = "UTF-8" ?>
<!DOCTYPE struts PUBLIC "-//Apache Software Foundation//DTD Struts Configuration 2.1//EN"
"http://struts.apache.org/dtds/struts-2.1.dtd">
<struts>
  <package name = "ch9" namespace = "/" extends = "struts-default">
    <action name = "upload" class = "action.UploadAction">
      <result>/uploadSuccess.jsp</result>
    </action>
  </package>
</struts>
```

配置了该 Action 后，如果在如图 9.1 所示的页面中输入账号，并浏览到需要上传的头像图片，然后单击"上传"按钮，该上传请求将被 UploadAction 处理，处理成功后转入 uploadSuccess.jsp 页面，该页面使用了简单的 Struts 2 标签来显示上传的头像。

uploadSuccess.jsp 页面的代码如下：

```jsp
<%@ page language = "java" import = "java.util.*" pageEncoding = "utf-8" %>
<%@ taglib prefix = "s" uri = "/struts-tags" %>
<html>
  <head>
    <title>My JSP 'uploadSuccess.jsp' starting page</title>
  </head>
  <body>
    上传成功！<br>
    <!-- 输出表单里的用户账号属性 -->
    用户账号：<s:property value = "uid"/><br>
    <!-- 根据上传的文件名，在页面显示上传的头像 -->
    您的头像：<img src = "<s:property value = "'images/' + headImageFileName"/>"/>
  </body>
</html>
```

如果上传成功，将看到如图 9.2 所示的页面。

### 3. Struts 2 多文件上传

在 Struts 2 应用中，如果一个页面有多个文件域需要实现上传，则可以为每个文件域提供 3 个属性，分别封装该文件域对应的文件名、文件类型和文件内容。多文件上传与单个文件上传并没有太大的区别。下面介绍使用数组处理同时上传多个文件的方式。

假设有如图 9.3 所示的上传文件页面。

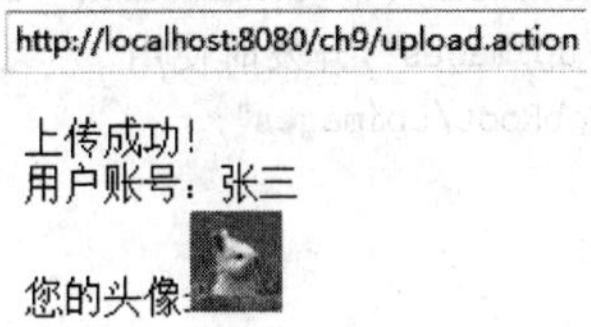

图9.2 上传成功页面 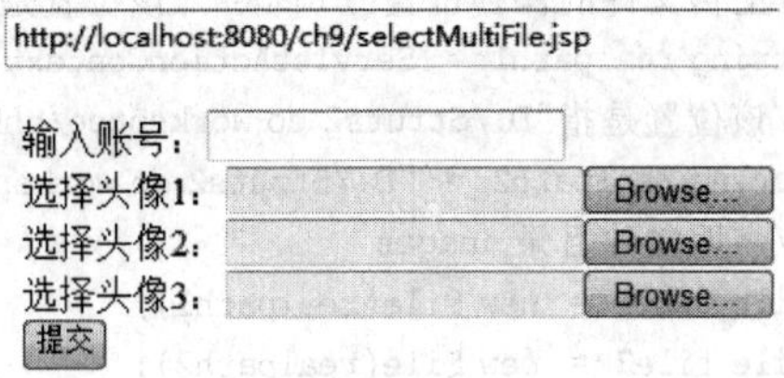　　　　图9.3 多文件上传页面

为了让数组一次封装3个文件域,需要将页面中3个文件域的 name 设置相同,下面是多文件上传页面代码。

```jsp
<%@ page language = "java" import = "java.util.*" pageEncoding = "utf-8" %>
<%@ taglib prefix = "s" uri = "/struts-tags" %>
<html>
  <head>
    <title>My JSP 'selectMultiFile.jsp' starting page</title>
  </head>
  <body>
    <s:form action = "uploadMulti.action" method = "post" theme = "simple" enctype = "multipart/form-data">
      输入账号:<s:textfield name = "uid"/><br>
      选择头像1:<s:file name = "headImage"/><br>
      选择头像2:<s:file name = "headImage"/><br>
      选择头像3:<s:file name = "headImage"/><br>
      <s:submit value = "提交"/>
    </s:form>
  </body>
</html>
```

在处理多文件上传的 Action 类中,需要使用3个数组分别封装文件名、文件类型和文件内容。下面是处理多文件上传的 Action 类的代码。

```java
package action;
import java.io.File;
import java.io.IOException;
import org.apache.commons.io.FileUtils;
import org.apache.struts2.ServletActionContext;
import com.opensymphony.xwork2.ActionSupport;
public class UploadMultiAction extends ActionSupport{
    private static final long serialVersionUID = 1L;
    //封装账号
    private String uid;
    //封装上传的文件
    private File[] headImage;
    //封装文件的类型
    private String[] headImageContentType;
    //封装文件的名称
    private String[] headImageFileName;
    public String execute() throws IOException{
```

```java
        //上传文件的保存位置"/images",该位置是指 tomcat 的 webapps\ch9\images
        String realpath1 = ServletActionContext.getServletContext().getRealPath("/images");
        //该位置是指"D:/Struts2 do workspace/ch9/WebRoot/upimages",开发时使用
        String realpath2 = "D:/Struts2 do workspace/ch9/WebRoot/upimages";
        //声明文件目录 images
        File file1 = new File(realpath1);
        File file2 = new File(realpath2);
        //如果目录不存在,则创建该目录
        if(!file1.exists()){
            file1.mkdirs();
        }
        //如果目录不存在,则创建该目录
        if(!file2.exists()){
            file2.mkdirs();
        }
        //实现多文件上传
        for( int i = 0 ;i< headImage.length; i++){
            File uploadImage = headImage[i];
            FileUtils.copyFile(uploadImage, new File(file1, headImageFileName[i]));
            FileUtils.copyFile(uploadImage, new File(file2, headImageFileName[i]));
        }
        return SUCCESS;
    }
    public String getUid() {
        return uid;
    }
    public void setUid(String uid) {
        this.uid = uid;
    }
    public File[] getHeadImage() {
        return headImage;
    }
    public void setHeadImage(File[] headImage) {
        this.headImage = headImage;
    }
    public String[] getHeadImageContentType() {
        return headImageContentType;
    }
    public void setHeadImageContentType(String[] headImageContentType) {
        this.headImageContentType = headImageContentType;
    }
    public String[] getHeadImageFileName() {
        return headImageFileName;
    }
    public void setHeadImageFileName(String[] headImageFileName) {
        this.headImageFileName = headImageFileName;
    }
}
```

配置多文件上传的 Action 与单文件上传的 Action 是一样的,具体代码如下。

```
<action name = "uploadMulti" class = "action.UploadMultiAction">
    <result>/uploadMultiSuccess.jsp</result>
</action>
```

多文件上传成功,将看到如图 9.4 所示的页面。

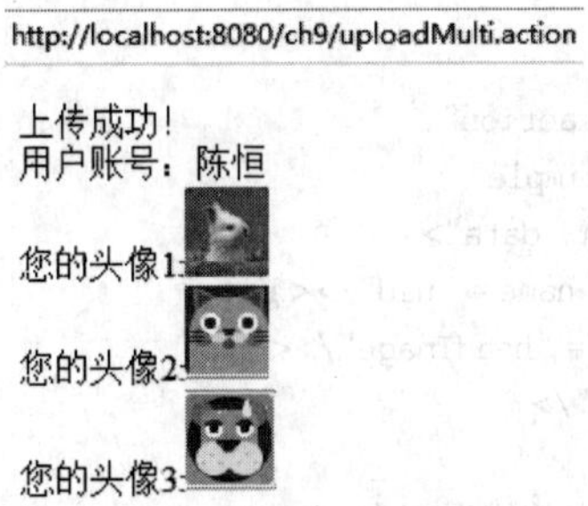

图 9.4  多文件上传成功页面

如图 9.4 所示页面(selectMultiFile.jsp)的具体代码如下:

```
<%@ page language = "java" import = "java.util.*" pageEncoding = "utf-8" %>
<%@ taglib prefix = "s" uri = "/struts-tags" %>
<html>
  <head>
    <title>My JSP 'uploadMultiSuccess.jsp' starting page</title>
  </head>
  <body>
    上传成功!<br>
    <!-- 输出表单里的用户账号属性 -->
    用户账号:<s:property value = "uid"/><br>
    <!-- 根据上传的文件名,在页面显示上传的头像 -->
    您的头像1:<img src = "<s:property value = "'images/' + headImageFileName[0]"/>"/><br>
    您的头像2:<img src = "<s:property value = "'images/' + headImageFileName[1]"/>"/><br>
    您的头像3:<img src = "<s:property value = "'images/' + headImageFileName[2]"/>"/><br>
  </body>
</html>
```

### 9.1.2  能力目标

掌握基于 Struts 2 框架文件上传的实现方式。

### 9.1.3  任务驱动

**1. 任务的主要内容**

修改 9.1.1 节中"2. Struts 2 单文件上传"示例的 Action 类,使得上传文件的扩展名限定为".jpg"".png"或"gif"。对应的 JSP 页面和 Action 配置文件都要做相应修改。

**2. 任务的代码模板**

selectFile.jsp 的代码模板如下:

```jsp
<%@ page language = "java" import = "java.util.*" pageEncoding = "utf-8"%>
<%@ taglib prefix = "s" uri = "/struts-tags" %>
<html>
  <head>
    <title>My JSP 'selectFile.jsp' starting page</title>
  </head>
  <body>
    <s:form action = "upload.action"
    method = "post" theme = "simple"
    enctype = "multipart/form-data">
     输入账号:<s:textfield name = "uid"/><br>
     选择头像:<s:file name = "headImage"/><br>
     <s:submit value = "提交"/>
    </s:form>
    <!-- 代码1 取出 Action 中的错误消息 -->
    【代码1】
  </body>
</html>
```

UploadAction.java 的代码模板如下:

```java
package action;
import java.io.File;
import java.io.IOException;
import org.apache.commons.io.FileUtils;
import org.apache.struts2.ServletActionContext;
import com.opensymphony.xwork2.ActionSupport;
public class UploadAction extends ActionSupport{
    private static final long serialVersionUID = 1L;
    //封装账号(uid)请求参数的属性
    private String uid;
    //封装上传文件域的属性
    private File headImage;
    //封装上传文件类型的属性
    private String headImageContentType;
    //封装上传文件名的属性
    private String headImageFileName;
    public String execute() throws IOException{
        //首先,判定文件名是否为".jpg"、".png"或".gif"
        String endName = headImageFileName.substring(headImageFileName.lastIndexOf("."));
        String patternName[] = {".jpg",".png",".gif"};
        boolean b = false;
        for (int i = 0; i < patternName.length; i++) {
            String string = patternName[i];
            if(string.equals(endName)){
                b = true;
                break;
            }
        }
        //文件名不符合规范
        if(!b){
```

```
        //代码2将错误消息"文件名不符合规范"以"filenameerror"为key,
        //存储在fieldErrors集合中.
        【代码2】
        return "input";
    }
    //上传文件的保存位置"/images",该位置是指tomcat的webapps\ch9\images,发布后使用
    String realpath1 = ServletActionContext.getServletContext().getRealPath("/images");
    //该位置是指"D:/Struts2 do workspace/ch9/WebRoot/upimages",开发时使用
    String realpath2 = "D:/Struts2 do workspace/ch9/WebRoot/upimages";
    //声明文件目录images
    File file1 = new File(realpath1);
    File file2 = new File(realpath2);
    //如果目录不存在,则创建该目录
    if(!file1.exists()){
          file1.mkdirs();
    }
    //如果目录不存在,则创建该目录
    if(!file2.exists()){
          file2.mkdirs();
    }
    //实现文件上传
    FileUtils.copyFile(headImage, new File(file1, headImageFileName));
    FileUtils.copyFile(headImage, new File(file2, headImageFileName));
    return SUCCESS;
}
public String getUid() {
    return uid;
}
public void setUid(String uid) {
    this.uid = uid;
}
public File getHeadImage() {
    return headImage;
}
public void setHeadImage(File headImage) {
    this.headImage = headImage;
}
public String getHeadImageContentType() {
    return headImageContentType;
}
public void setHeadImageContentType(String headImageContentType) {
    this.headImageContentType = headImageContentType;
}
public String getHeadImageFileName() {
    return headImageFileName;
}
public void setHeadImageFileName(String headImageFileName) {
    this.headImageFileName = headImageFileName;
}
}
```

配置文件的代码模板如下：

```xml
<action name="upload" class="action.UploadAction">
    <result>/uploadSuccess.jsp</result>
    <!-- 代码3 配置input视图(selectFile.jsp) -->
    【代码3】
</action>
```

### 3. 任务小结或知识扩展

**1) 拦截器实现文件过滤**

Struts 2 提供了一个名为 fileUpload 的拦截器，通过配置该拦截器可以更轻松地实现文件过滤。为了让 fileUpload 拦截器起作用，只需要在处理文件上传的 Action 中配置该拦截器引用即可。

配置 fileUpload 拦截器时可以指定如下两个参数。

① allowedTypes：该参数指定允许上传的文件类型，多个文件类型之间以英文逗号(,)隔开。

② maximumSize：该参数指定允许上传的文件大小，单位是字节。

当文件过滤失败后，系统自动转入 input 逻辑视图，因此必须为 Action 配置名为 input 的逻辑视图。

通过拦截器来实现文件过滤的配置文件示例如下：

```xml
<?xml version="1.0" encoding="UTF-8"?>
<!DOCTYPE struts PUBLIC "-//Apache Software Foundation//DTD Struts Configuration 2.1//EN" "http://struts.apache.org/dtds/struts-2.1.dtd">
<struts>
 <!-- 指定国际化资源文件的 baseName 为 messageResource -->
 <constant name="struts.custom.i18n.resources" value="messageResource"/>
 <package name="ch9" namespace="/" extends="struts-default">
     <interceptors>
         <!-- 配置拦截器栈 -->
         <interceptor-stack name="myStack">
             <!-- 配置 fileUpload 拦截器 -->
             <interceptor-ref name="fileUpload">
                 <!-- 配置允许上传的文件类型,需要注意的是不同的浏览器,文件类型的写法可能不一样 -->
                 <param name="allowedTypes">
                     image/x-png,image/bmp,image/png,image/gif,image/jpeg,image/jpg
                 </param>
                 <!-- 配置允许上传文件的大小,单位是字节 -->
                 <param name="maximumSize">2097152</param>
             </interceptor-ref>
             <interceptor-ref name="defaultStack"/>
         </interceptor-stack>
     </interceptors>
     <action name="upload" class="action.UploadAction">
         <!-- 使用拦截器栈 -->
         <interceptor-ref name="myStack"/>
```

```
            <result>/uploadSuccess.jsp</result>
            <result name="input">/selectFile.jsp</result>
        </action>
    </package>
</struts>
```

如果上传失败,系统应回到文件上传页面,并显示错误信息。因此,需要在文件上传页面 selectFile.jsp 中加上"<s:fielderror/>"。

如果上传的文件大于 2MB,将看到如图 9.5 所示的错误提示页面。

图 9.5　文件太大提示页面

图 9.5 的提示信息是系统默认的提示信息,所以应该使用国际化信息替换它。

在 Struts 2 中上传文件太大的提示信息的 key 是"struts.messages.error.file.too.large",开发者可以在自己的国际化资源文件中增加该 key 的信息,即可改变该提示信息。

同理,在 Struts 2 中不允许上传的文件类型提示信息的 key 是"struts.messages.error.content.type.not.allowed",开发者也可以在自己的国际化资源文件中增加该 key 的信息,即可改变文件类型不允许的提示信息。

2) 文件上传的常量配置

上传文件时,系统默认使用 Web 服务器的工作路径作为临时路径。为了避免文件上传时使用 Web 服务器的工作路径作为临时路径,则应该设置 struts.multipart.saveDir 常量。该常量指定上传文件的临时保存路径。该常量配置示例如下:

```
<constant name="struts.multipart.saveDir" value="d:/tmpuploadfiles"/>
```

除此之外,还有一个文件上传的常量 struts.multipart.maxSize。该常量指定 Struts 2 文件上传中整个请求内容所允许的最大字节数,默认为 2097152(2MB)。该常量配置示例如下:

```
<constant name="struts.multipart.maxSize" value="20971520"/>
```

**4. 任务代码模板的参考答案**

【代码 1】<s:fielderror/>

【代码 2】this.addFieldError("filenameerror", "文件名不符合规范");

【代码 3】<result name="input">/selectFile.jsp</result>

## 9.1.4　实践环节

使用 fileUpload 拦截器将 9.1.1 节中"3. Struts 2 多文件上传"的文件扩展名限定为".jpg"".png"或"gif"。

## 9.2 Struts 2 文件下载

很多读者认为文件下载很简单,直接在页面上给出一个链接,该链接的 href 属性等于要下载文件的文件名,不就可以实现文件下载了吗?大多数情况下,确实可以实现文件下载,但如果被下载文件的文件名为中文文件名,则会导致下载失败。另外,有时候应用程序需要在用户下载之前进行进一步检查,比如判断用户是否有足够权限来下载该文件等,可以使用 Struts 2 的文件下载支持来解决这些问题。下面通过一个简单示例讲解 Struts 2 文件下载的使用方法。

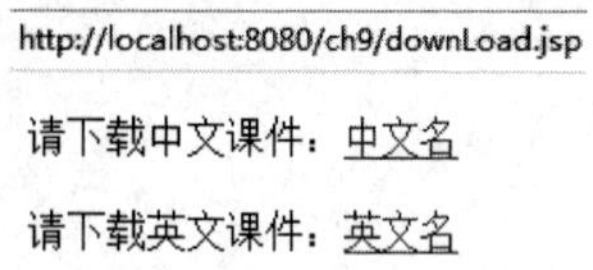

假设有如图 9.6 所示的下载页面。

请下载中文课件:中文名

请下载英文课件:英文名

图 9.6 文件下载页面

如图 9.6 所示,有两个文件可下载,一个是中文的文件名,一个是英文的文件名。该页面的具体代码如下:

```
<%@ page language="java" import="java.util.*" pageEncoding="utf-8"%>
<html>
  <head>
    <title>My JSP 'downLoad.jsp' starting page</title>
  </head>
  <body>
      <!-- 在 webapps\ch9\images 目录下有文件"第 11 章.doc"和"ch11.doc" -->
      请下载中文课件:<a href="downLoad.action?downPath=第 11 章.doc">中文名</a><br><br>
      请下载英文课件:<a href="downLoad.action?downPath=ch11.doc">英文名</a><br>
  </body>
</html>
```

在 Struts 2 的文件下载 Action 类中,需要提供一个返回 InputStream 流的方法,该输入流代表了被下载文件的入口。该 Action 类的代码如下:

```
package action;
import java.io.InputStream;
import java.io.UnsupportedEncodingException;
import org.apache.struts2.ServletActionContext;
import util.MyUtil;
import com.opensymphony.xwork2.ActionSupport;
public class DownLoadAction extends ActionSupport {
    // 下载时的文件名
    private String downPath;
    // 保存文件的类型
    private String contentType;
    // 保存时的文件名
    private String filename;
    public String getContentType() {
        return contentType;
    }
    public void setContentType(String contentType) {
        this.contentType = contentType;
```

```java
    }
    public String getFilename() {
        return filename;
    }
    public void setFilename(String filename) {
        this.filename = filename;
    }
    public String getDownPath() {
        return downPath;
    }
    public void setDownPath(String downPath) {
        try {
            // 解决被下载文件的中文文件名乱码问题
            downPath = new String(downPath.getBytes("ISO-8859-1"), "UTF-8");
        } catch (UnsupportedEncodingException e) {
            e.printStackTrace();
        }
        this.downPath = downPath;
    }
    /*
     * 下载用的 Action 返回一个 InputStream 实例,该方法对应 Action 配置里的
     * result 的 inputName 参数,值为 inputStream
     */
    public InputStream getInputStream() {
        return ServletActionContext.getServletContext().getResourceAsStream(
                downPath);
    }
    public String execute() {
        // 下载保存时的文件名和被下载的文件名一样
        filename = downPath;
        // 下载的文件路径,请在 webapps\ch9 目录下创建 images
        downPath = "images/" + downPath;
        // 保存文件的类型
        contentType = "application/x-msdownload";
        // 对下载的文件名按照 UTF-8 进行编码,解决下载窗口中的中文乱码问题
        filename = MyUtil.toUTF8String(filename);
        return SUCCESS;
    }
}
```

上述 Action 类中使用了一个自定义工具类 MyUtil,该类中有个静态方法 toUTF8String 实现对下载的文件名按照 UTF-8 进行编码,解决下载窗口中的中文乱码问题。具体代码如下:

```java
package util;
import java.io.UnsupportedEncodingException;
public class MyUtil {
    //对下载的文件名按照 UTF-8 进行编码
    public static String toUTF8String(String str) {
        StringBuffer sb = new StringBuffer();
        int len = str.length();
```

```java
            for (int i = 0; i < len; i++) {
                // 取出字符中的每个字符
                char c = str.charAt(i);
                // Unicode 码值在 0～255,不作处理
                if (c >= 0 && c <= 255) {
                    sb.append(c);
                } else {// 转换 UTF-8 编码
                    byte b[];
                    try {
                        b = Character.toString(c).getBytes("UTF-8");
                    } catch (UnsupportedEncodingException e) {
                        e.printStackTrace();
                        b = null;
                    }
                    // 转换为 %HH 的字符串形式
                    for (int j = 0; j < b.length; j++) {
                        int k = b[j];
                        if (k < 0) {
                            k &= 255;
                        }
                        sb.append("%" + Integer.toHexString(k).toUpperCase());
                    }
                }
            }
            return sb.toString();
        }
    }
```

配置文件下载的 Action 与配置普通的 Action 基本是一样的,关键是需要配置一个类型为 stream 的结果(result),配置时需要指定如下 4 个属性。

(1) contentType：指定被下载文件的文件类型。

(2) inputName：指定被下载文件的入口输入流。

(3) contentDisposition：指定下载的文件名。

(4) bufferSize：指定下载文件时的缓冲大小。

stream 类型的 result 的逻辑视图是返回给客户端一个输入流,因此无须指定位置。具体配置代码如下：

```xml
<action name="downLoad" class="action.DownLoadAction">
    <!-- 结果类型为 stream -->
    <result type="stream">
        <!-- 指定下载文件的文件类型,默认为 text/plain -->
        <param name="contentType">${contentType}</param>
        <!-- 默认就是 inputStream,它将会指示 StreamResult 通过 inputName 属性值的 getter
        方法(比如这里就是 getInputStream())来获取下载文件的内容,意味着 Action 要有这个方法 -->
        <param name="inputName">inputStream</param>
        <!-- 默认为 inline(在线打开),设置为 attachment 将会告诉浏览器下载该文件,
        filename 指定下载文件时的文件名,若未指定将会以浏览的页面名为文件名,如以 download.action
        为文件名 -->
        <param name="contentDisposition">attachment;filename=${filename}</param>
```

```
            <!-- 指定下载文件的缓冲大小 -->
            <param name = "bufferSize">4096</param>
        </result>
</action>
```

单击图 9.6 中的"中文名"链接,打开如图 9.7 所示的下载框。

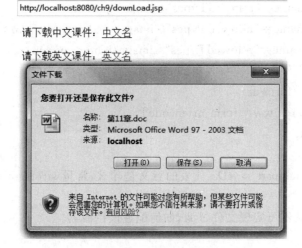

图 9.7　中文文件名文件下载

单击图 9.6 中"英文名"链接,打开如图 9.8 所示的下载框。

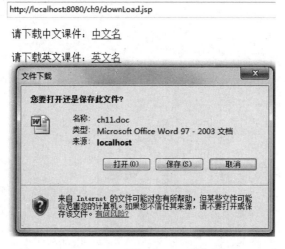

图 9.8　英文文件名文件下载

## 9.3　本章小结

本章重点介绍了 Struts 2 的文件上传,主要包括如何使用 3 个属性封装文件域的文件名、文件类型和文件内容。详细介绍了如何使用拦截器过滤上传文件的类型、大小。最后介绍了如何利用 stream 类型的结果控制文件下载。

# 习 题 9

1. 以下配置文件上传拦截器,只允许 bmp 类型的图片上传,正确的是(　　)。
   A. <param name="allowedTypes">*.bmp</param>
   B. <param name="allowedTypes">bmp</param>
   C. <param name="allowedTypes">image/*.bmp</param>
   D. <param name="allowedTypes">image/bmp</param>
2. 基于表单的文件上传,应将表单的 enctype 属性值设置为(　　)。
   A. multipart/form-data
   B. application/x-www-form-urlencoded
   C. text/plain
   D. html/text
3. 常量 struts.multipart.saveDir 代表的意义是什么,常量 struts.multipart.maxSize 代表的意义又是什么?
4. 哪个内置拦截器可实现文件过滤,如何配置该拦截器?

# 第10章 名片管理系统的设计与实现

**主要内容**

(1) 系统设计。
(2) 数据库设计。
(3) 系统管理。
(4) 组件设计。
(5) 系统实现。

本章系统使用 Struts 2 框架实现各个模块,Web 引擎为 Tomcat 7.0,数据库采用的是 MySQL 5.x,集成开发环境为 MyEclipse Professional 2014。

## 10.1 系统设计

### 10.1.1 系统功能需求

名片管理系统是针对注册用户使用的系统,系统提供的功能如下。
(1) 非注册用户可以注册为注册用户。
(2) 成功注册的用户,可以登录系统。
(3) 成功登录的用户,可以添加、修改、删除以及浏览自己客户的名片信息。
(4) 成功登录的用户,可以在个人中心查看自己的基本信息和修改密码。

### 10.1.2 系统模块划分

用户登录成功后,进入管理主页面(main.jsp)可以对自己的客户名片进行管理。系统模块划分,如图 10.1 所示。

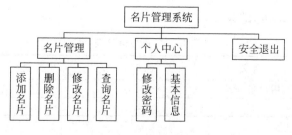

图 10.1 名片管理系统

## 10.2 数据库设计

系统采用加载纯 Java 数据库驱动程序的方式连接 MySQL 5.x 数据库。在 MySQL 5.x 的数据库 card 中,共创建两张与系统相关的数据表:usertable 和 cardinfo。

### 10.2.1 数据库概念结构设计

根据系统设计与分析,可以设计出如下数据结构。

**1. 用户**

用户包括用户名和密码,注册用户名唯一。

**2. 名片**

名片包括 ID、名称、电话、邮箱、单位、职务、地址、logo 以及所属用户。其中,ID 唯一,"所属用户"与"1.用户"关联。

根据以上的数据结构,结合数据库设计的特点,可画出如图 10.2 所示的数据库概念结构图。

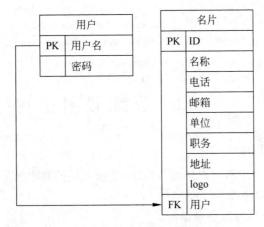

图 10.2 数据库概念结构图

其中,ID 为正整数,值是从 1 开始递增的序列。

### 10.2.2 数据库逻辑结构设计

将数据库概念结构图转换为 MySQL 数据库所支持的实际数据模型,即数据库的逻辑结构。

用户信息表(usertable)的设计,如表 10.1 所示。

表 10.1 用户信息表

| 字段 | 含义 | 类型 | 长度 | 是否为空 |
| --- | --- | --- | --- | --- |
| userName | 用户名(PK) | varchar | 50 | no |
| password | 密码 | varchar | 20 | no |

名片信息表(cardinfo)的设计,如表 10.2 所示。

表 10.2　名片信息表

| 字段 | 含义 | 类型 | 长度 | 是否为空 |
| --- | --- | --- | --- | --- |
| id | 编号(PK) | int | 11 | no |
| name | 名称 | varchar | 50 | no |
| telephone | 电话 | varchar | 20 | no |
| email | 邮箱 | varchar | 20 | |
| company | 单位 | varchar | 50 | |
| post | 职务 | varchar | 50 | |
| address | 地址 | varchar | 50 | |
| logo | 图片 | varchar | 25 | |
| userName | 所属用户 | varchar | 50 | no |

## 10.3　系统管理

### 10.3.1　导入相关的 jar 包

新建一个 Struts 2 应用 cardManage,在所有 JSP 页面中尽量使用 Struts 2 标签、EL 表达式和 JSTL 标签,又因为系统采用纯 Java 数据库驱动程序连接 MySQL 5.x。所以,需要将对应的驱动程序"mysql-connector-java-5.0.6-bin.jar"复制到 cardManage/ WebRoot/ WEB-INF/lib 的目录下。

### 10.3.2　JSP 页面管理

由于篇幅受限,本章仅附上 JSP 和 Java 文件的代码。

#### 1. 管理主页面

注册用户在浏览器地址栏中输入 http://localhost:8080/cardManage/login.jsp 访问登录页面,登录成功后,进入管理主页面(main.jsp),main.jsp 的运行效果如图 10.3 所示。

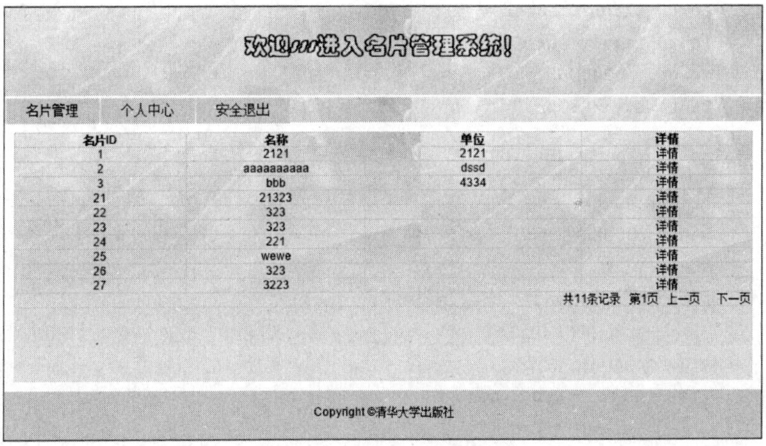

图 10.3　管理主页面

管理主页面 main.jsp 的代码如下：

```jsp
<%@ page language="java" contentType="text/html; charset=UTF-8" pageEncoding="UTF-8"%>
<%@taglib prefix="s" uri="/struts-tags" %>
<%
    String path = request.getContextPath();
    String basePath = request.getScheme()+"://"+request.getServerName()+":"+request.getServerPort()+path+"/";
%>
<!DOCTYPE html PUBLIC "-//W3C//DTD HTML 4.01 Transitional//EN" "http://www.w3.org/TR/html4/loose.dtd">
<html>
    <head>
        <base href="<%=basePath%>">
        <title>后台主页面</title>
        <style type="text/css">
            * {
                margin: 0px;
                padding: 0px;
            }
            body {
                font-family: Arial, Helvetica, sans-serif;
                font-size: 12px;
                margin: 0px auto;
                height: auto;
                width: 800px;
                border: 1px solid #006633;
            }
            #header {
                height: 90px;
                width: 800px;
                background-image: url(images/bb.jpg);
                margin: 0px 0px 3px 0px;
            }
            #header h1 {
                text-align: center;
                font-family: 华文彩云;
                color: #000000;
                font-size: 30px;
            }
            #navigator {
                height: 25px;
                width: 800px;
                font-size: 14px;
                background-image: url(images/bb.jpg);
            }
            #navigator ul {
                list-style-type: none;
            }
            #navigator li {
```

```css
            float: left;
            position: relative;
        }
        #navigator li a {
            color: #000000;
            text-decoration: none;
            padding-top: 4px;
            display: block;
            width: 98px;
            height: 22px;
            text-align: center;
            background-color: PaleGreen;
            margin-left: 2px;
        }
        #navigator li a:hover {
            background-color: #006633;
            color: #FFFFFF;
        }
        #navigator ul li ul {
           visibility: hidden;
           position: absolute;
        }
        #navigator ul li:hover ul,
        #navigator ul a:hover ul{
           visibility: visible;
        }
        #content {
            height: auto;
            width: 780px;
            padding: 10px;
        }
        #content iframe {
            height: 300px;
            width: 780px;
        }
        #footer {
            height: 30px;
            width: 780px;
            line-height: 2em;
            text-align: center;
            background-color: PaleGreen;
            padding: 10px;
        }
    }
    </style>
</head>
<body>
    <div id="header">
        <br>
        <br>
        <h1>欢迎<s:property value="#session.userName"/>进入名片管理系统!</h1>
    </div>
```

```html
        <div id="navigator">
            <ul>
                <li><a>名片管理</a>
                    <ul>
                        <li><a href="addCard.jsp" target="center">添加名片</a></li>
                        <li><a href="card/queryCard.action?act=deleteSelect" target="center">删除名片</a></li>
                        <li><a href="card/queryCard.action?act=updateSelect" target="center">修改名片</a></li>
                        <li><a href="card/queryCard.action" target="center">查询名片</a></li>
                    </ul>
                </li>
                <li><a>个人中心</a>
                    <ul>
                        <li><a href="updatePWD.jsp" target="_top">修改密码</a></li>
                        <li><a href="userInfo.jsp" target="center">基本信息</a></li>
                    </ul>
                </li>
                <li><a href="user/exit.action">安全退出</a></li>
            </ul>
        </div>
        <div id="content">
            <iframe src="card/queryCard.action" name="center" frameborder="0"></iframe>
        </div>
        <div id="footer">Copyright ©清华大学出版社</div>
    </body>
</html>
```

### 2. 程序报错页面

当Java程序运行出现异常时,系统会执行全局页面error.jsp,具体代码如下:

```html
<body>
    程序内部错误。<br>
</body>
```

### 3. 无权限提示页面

在没有成功登录的情况下,对名片进行增、删、改、查等操作时,系统执行无权限操作提示页面nologin.jsp,具体代码如下:

```jsp
<%@ page language="java" import="java.util.*" pageEncoding="UTF-8"%>
<%
    String path = request.getContextPath();
    String basePath = request.getScheme()+"://"+request.getServerName()+":"+request.getServerPort()+path+"/";
%>
<!DOCTYPE HTML PUBLIC "-//W3C//DTD HTML 4.01 Transitional//EN">
<html>
    <head>
        <base href="<%=basePath%>">
```

```
        <title>My JSP 'nologin.jsp' starting page</title>
    </head>
    <body>
        您没有登录成功,无权操作。请<a href = "login.jsp" target = "_top">登录</a>!
    </body>
</html>
```

### 10.3.3 包管理

本系统的包层次结构如图 10.4 所示。

#### 1. action 包

该包包括系统中所有的 Action 类,包括"名片管理"的 Action 类和"个人中心"的 Action 类。

#### 2. conf 包

该包的 xml 文件是本系统所有 Action 类的配置。card.xml 是"名片管理"Action 类的配置,user.xml 是"个人中心"Action 类的配置。这些 xml 文件需要在 struts.xml 文件中包含进来。

#### 3. dao 包

dao 包中存放的 Java 程序实现数据库的操作。其中 BaseDao 是一个父类,该类负责连接数据库;CardDao 是 BaseDao 的一个子类,有关"名片管理"的数据访问在该类中;UserDao 是 BaseDao 的另一个子类,有关用户的数据访问在该类中。

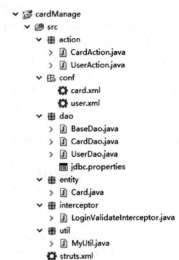

图 10.4 包层次结构图

另外,该包中还有一个名为 jdbc.properties 的文件,该文件是有关数据库的配置。包括驱动类名、数据库 URL、用户名以及密码等。

#### 4. entity 包

该包中有 1 个实体类:Card,封装名片信息。

#### 5. interceptor 包

该包中有 1 个拦截器类:LoginValidateInterceptor,进行权限控制。

#### 6. util 包

该包中 MyUtil 类是获得一个时间字符串的工具类。

### 10.3.4 配置文件管理

#### 1. struts.xml

在软件系统开发中,模块化设计是最常用的一种方式。为了方便管理,本系统也不例外,将不同模块的配置分别放在不同的配置文件中。然后,在 struts.xml 文件中,通过 include 引入这些配置文件。

struts.xml 文件的配置代码如下:

```xml
<?xml version="1.0" encoding="UTF-8"?>
<!DOCTYPE struts PUBLIC "-//Apache Software Foundation//DTD Struts Configuration 2.1//EN"
"http://struts.apache.org/dtds/struts-2.1.dtd">
<struts>
    <constant name="struts.multipart.saveDir" value="d:/tmpuploadfiles"/>
    <constant name="struts.multipart.maxSize" value="20971520"/>
    <!-- 设置表单主题为simple -->
    <constant name="struts.ui.theme" value="simple"/>
    <package name="mydefault" namespace="/mydefault" extends="struts-default">
        <interceptors>
            <!-- 定义拦截器loginInterceptor -->
            <interceptor name="loginInterceptor" class="interceptor.LoginValidateInterceptor">
            </interceptor>
            <!-- 定义拦截器栈myStack -->
            <interceptor-stack name="myStack">
                <!-- 使用拦截器loginInterceptor -->
                <interceptor-ref name="loginInterceptor"></interceptor-ref>
                <!-- 使用内置拦截器 -->
                <interceptor-ref name="defaultStack"></interceptor-ref>
            </interceptor-stack>
        </interceptors>
        <global-results>
            <!-- 全局转发,程序出错时,也可使用全局异常 -->
            <result name="error">/error.jsp</result>
            <result name="nologin">/nologin.jsp</result>
        </global-results>
    </package>
    <include file="conf/user.xml"></include>
    <include file="conf/card.xml"></include>
</struts>
```

## 2. user.xml

该配置文件负责配置和用户有关的Action,具体配置代码如下:

```xml
<?xml version="1.0" encoding="UTF-8"?>
<!DOCTYPE struts PUBLIC "-//Apache Software Foundation//DTD Struts Configuration 2.1//EN"
"http://struts.apache.org/dtds/struts-2.1.dtd">
<struts>
    <package name="user" namespace="/user" extends="mydefault">
        <action name="regist" class="action.UserAction" method="register">
            <result name="register">/register.jsp</result>
            <result>/login.jsp</result>
        </action>
        <action name="login" class="action.UserAction" method="login">
            <result name="loginFail">/login.jsp</result>
            <result>/main.jsp</result>
        </action>
        <action name="exit" class="action.UserAction" method="logout">
            <result>/login.jsp</result>
        </action>
```

```xml
<action name = "updatePwd" class = "action.UserAction" method = "updatePwd">
    <result>/login.jsp</result>
</action>
    </package>
</struts>
```

### 3. card.xml

该配置文件负责配置个人中心有关的 Action，在此处使用了通配符"*"动态匹配 Action 的请求名称，具体配置代码如下：

```xml
<?xml version = "1.0" encoding = "UTF-8"?>
<!DOCTYPE struts PUBLIC " -//Apache Software Foundation//DTD Struts Configuration 2.1//EN"
"http://struts.apache.org/dtds/struts-2.1.dtd">
<struts>
    <package name = "card" namespace = "/card" extends = "mydefault">
        <action name = "*Card" class = "action.CardAction" method = "{1}">
            <!-- 使用拦截器进行登录权限验证 -->
            <interceptor-ref name = "myStack"></interceptor-ref>
            <result name = "addSuccess" type = "redirect">/card/queryCard.action</result>
            <result name = "addFail">/addCard.jsp</result>
            <result name = "querySuccess">/queryCards.jsp</result>
            <result name = "detailcard">/detail.jsp</result>
            <result name = "deleteSelect">/deleteSelect.jsp</result>
            <result name = "deleteSucess" type = "redirect">/card/queryCard.action?act = deleteSelect</result>
            <result name = "updateSelect">/updateSelect.jsp</result>
            <result name = "updateAcard">/updateCard.jsp</result>
            <result name = "updateSuccess" type = "redirect">/card/queryCard.action?act = updateSelect</result>
        </action>
    </package>
</struts>
```

## 10.4 组件设计

本系统的组件包括工具类、拦截器类和数据库操作(dao)。

### 10.4.1 工具类

工具类 MyUtil 的代码如下：

```java
package util;
import java.text.SimpleDateFormat;
import java.util.Date;
public class MyUtil {
    /**
     * 获得一个以时间字符串为标准的 ID,固定长度是 17 位
     */
    public static String getStringID(){
```

```java
            String id = null;
            Date date = new Date();
            SimpleDateFormat sdf = new SimpleDateFormat("yyyyMMddHHmmssSSS");
            id = sdf.format(date);
            return id;
        }
    }
```

### 10.4.2 拦截器类

拦截器类 LoginValidateInterceptor 的代码如下：

```java
package interceptor;
import java.util.Map;
import com.opensymphony.xwork2.ActionContext;
import com.opensymphony.xwork2.ActionInvocation;
import com.opensymphony.xwork2.interceptor.Interceptor;
public class LoginValidateInterceptor implements Interceptor{
    private static final long serialVersionUID = 1L;
    @Override
    public void destroy() {
    }
    @Override
    public void init() {
    }
    @Override
    public String intercept(ActionInvocation arg0) throws Exception {
            // 通过核心调度器 invocation 来获得调度的 Action 上下文
            ActionContext actionContext = arg0.getInvocationContext();
            //获得 session 对象
            Map<String,Object> session = actionContext.getSession();
            //从 session 中取出名为 user 的 session 属性
            String user = (String)session.get("userName");
            if(user != null ){
                return arg0.invoke();
            }else{
                //返回全局视图"nologin"
                return "nologin";
            }
    }
}
```

### 10.4.3 数据库操作

本系统有关数据库操作的 Java 类位于包 dao 中。为了方便管理，有关"名片管理"的数据访问由 CardDao 实现；有关"个人中心"的数据访问由 UserDao 实现。

（1）包 dao 中的配置文件 jdbc.properties 的内容如下：

```
jdbc.driverClass = com.mysql.jdbc.Driver
jdbc.jdbcUrl = jdbc:mysql://localhost:3306/card
```

```
jdbc.username = root
jdbc.password = root
```

(2) BaseDao.java 的代码如下:

```java
package dao;
import java.sql.Connection;
import java.sql.DriverManager;
import java.sql.PreparedStatement;
import java.sql.ResultSet;
import java.util.ArrayList;
import java.util.Properties;
public class BaseDao {
    // list 连接池
    static ArrayList<Connection> list = new ArrayList<Connection>();
    /**
     * 从连接池中获得一个连接
     */
    public synchronized static Connection getConnection() throws Exception{
        Connection con = null;
        // 如果连接池有连接
        if (list.size() > 0) {
            return list.remove(0);
        }
        // 连接池中没有连接
        else {
            Properties p = new Properties();
            //加载配置文件
            p.load(BaseDao.class.getClassLoader().getResourceAsStream("dao/jdbc.properties"));
            String driverClass = p.getProperty("jdbc.driverClass");
            String jdbcUrl = p.getProperty("jdbc.jdbcUrl");
            String username = p.getProperty("jdbc.username");
            String password = p.getProperty("jdbc.password");
            //加载驱动
            Class.forName(driverClass);
            // 和指定的数据库建立连接
            for (int i = 0; i < 10; i++) {
                con = DriverManager.getConnection(jdbcUrl, username, password);
                list.add(con);
            }
        }
        return list.remove(0);
    }
    /**
     * 关闭结果集
     * @param rs 代表结果集
     */
    public static void close(ResultSet rs) throws Exception{
        if (rs != null)
            rs.close();
    }
```

```java
/**
 * 关闭预处理对象
 * @param pst 代表预处理
 */
public static void close(PreparedStatement pst) throws Exception{
    if (pst != null)
        pst.close();
}
/**
 * 关闭连接对象
 * @param con 代表连接对象
 */
public synchronized static void close(Connection con) {
    if (con != null)
        list.add(con);
}
/**
 * 关闭结果集、预处理、连接等对象
 * @param rs 结果集
 * @param ps 预处理
 * @param con 连接
 */
public static void close(ResultSet rs, PreparedStatement ps, Connection con) throws Exception{
    close(rs);
    close(ps);
    close(con);
}
}
```

(3) CardDao.java 的代码如下：

```java
package dao;
import java.sql.Connection;
import java.sql.PreparedStatement;
import java.sql.ResultSet;
import java.util.ArrayList;
import entity.Card;
public class CardDao extends BaseDao{
    /**
     * 添加名片信息
     */
    public boolean add(Card card) throws Exception{
        boolean b = false;
        Connection con = getConnection();
        //其中 ID 为自增
        String sql = "insert into cardinfo values(null,?,?,?,?,?,?,?,?)";
        PreparedStatement ps = con.prepareStatement(sql);
        ps.setString(1, card.getName());
        ps.setString(2, card.getTelephone());
        ps.setString(3, card.getEmail());
        ps.setString(4, card.getCompany());
```

```java
            ps.setString(5, card.getPost());
            ps.setString(6, card.getAddress());
            ps.setString(7, card.getNewFileName());
            ps.setString(8, card.getUserName());
            int i = ps.executeUpdate();
            if(i > 0)
                b = true;
            return b;
    }
    /**
     * 修改名片信息
     */
    public boolean update(Card card) throws Exception{
            boolean b = false;
            Connection con = getConnection();
            String sql = "update cardinfo set name = ?, "
                    + " telephone = ?, "
                    + " email = ?, "
                    + " company = ?, "
                    + " post = ?, "
                    + " address = ? ";
            if(card.getNewFileName() != null){
                sql = sql + ", logo = ? ";
            }
            sql = sql + " where id = ? ";
            PreparedStatement ps = con.prepareStatement(sql);
            ps.setString(1, card.getName());
            ps.setString(2, card.getTelephone());
            ps.setString(3, card.getEmail());
            ps.setString(4, card.getCompany());
            ps.setString(5, card.getPost());
            ps.setString(6, card.getAddress());
            if(card.getNewFileName() != null){
                ps.setString(7, card.getNewFileName());
                ps.setString(8, card.getId());
            }else{
                ps.setString(7, card.getId());
            }
            int i = ps.executeUpdate();
            if(i > 0)
                b = true;
            return b;
    }
    /**
     * 查询所有名片
     */
    public ArrayList<Card> queryAll(String userName) throws Exception{
            ArrayList<Card> ac = new ArrayList<Card>();
            Connection con = getConnection();
            String sql = " select * from cardinfo where userName = ? ";
            PreparedStatement ps = con.prepareStatement(sql);
```

```java
            ps.setString(1,userName);
            ResultSet rs = ps.executeQuery();
            while(rs.next()){
                Card c = new Card();
                c.setId(rs.getString("id"));
                c.setName(rs.getString("name"));
                c.setTelephone(rs.getString("telephone"));
                c.setEmail(rs.getString("email"));
                c.setCompany(rs.getString("company"));
                c.setPost(rs.getString("post"));
                c.setAddress(rs.getString("address"));
                c.setNewFileName(rs.getString("logo"));
                ac.add(c);
            }
            return ac;
    }
    /**
     * 分页查询
     */
    public ArrayList<Card> queryByPage(int pageCur, String userName) throws Exception {
            ArrayList<Card> ac = new ArrayList<Card>();
            Connection con = getConnection();
            String sql = "select a.id,a.name,"
                    + " a.telephone,a.email, a.company, a.post, a.address, a.logo "
                    + " from cardinfo a where a.userName=? limit ?,? ";
            PreparedStatement ps = con.prepareStatement(sql);
            ps.setString(1,userName);
            ps.setInt(2,(pageCur-1)*10);
            ps.setInt(3,pageCur*10);
            ResultSet rs = ps.executeQuery();
            while(rs.next()){
                Card c = new Card();
                c.setId(rs.getString("id"));
                c.setName(rs.getString("name"));
                c.setTelephone(rs.getString("telephone"));
                c.setEmail(rs.getString("email"));
                c.setCompany(rs.getString("company"));
                c.setPost(rs.getString("post"));
                c.setAddress(rs.getString("address"));
                c.setNewFileName(rs.getString("logo"));
                ac.add(c);
            }
            return ac;
    }
    /**
     * 查询一个名片
     * @throws Exception
     */
    public Card selectA(String id) throws Exception{
            Card c = new Card();
            Connection con = getConnection();
```

```java
            String sql = " select * from cardinfo where id = ? ";
            PreparedStatement ps = con.prepareStatement(sql);
            ps.setString(1, id);
            ResultSet rs = ps.executeQuery();
            if(rs.next()){
                c.setId(rs.getString("id"));
                c.setName(rs.getString("name"));
                c.setTelephone(rs.getString("telephone"));
                c.setEmail(rs.getString("email"));
                c.setCompany(rs.getString("company"));
                c.setPost(rs.getString("post"));
                c.setAddress(rs.getString("address"));
                c.setNewFileName(rs.getString("logo"));
            }
            return c;
        }
        /**
         * 删除一个名片信息
         */
        public boolean delete(String id) throws Exception{
            boolean b = false;
            Connection con = getConnection();
            String sql = "delete from cardinfo where id = ?";
            PreparedStatement ps = con.prepareStatement(sql);
            ps.setString(1, id);
            int i = ps.executeUpdate();
            if(i > 0)
                b = true;
            return b;
        }
        /**
         * 删除多个名片信息
         */
        public boolean delete(String ids[]) throws Exception{
            boolean b = false;
            Connection con = getConnection();
            String sql = "delete from cardinfo where id in ( ";
            for (int i = 0; i < ids.length - 1; i++) {
                sql = sql + ids[i] + ",";
            }
            sql = sql + ids[ids.length - 1] + " )";
            PreparedStatement ps = con.prepareStatement(sql);
            int i = ps.executeUpdate();
            if(i > 0)
                b = true;
            return b;
        }
}
```

(4) UserDao.java 的代码如下：

```java
package dao;
import java.sql.Connection;
import java.sql.PreparedStatement;
import java.sql.ResultSet;
public class UserDao extends BaseDao{
    /**
     * 判断用户名是否存在
     * @throws Exception
     */
    public boolean isExit(String uname) throws Exception{
        boolean b = false;          //用户名不存在
        Connection con = getConnection();
        String sql = "select * from usertable where userName = ?";
        PreparedStatement ps = con.prepareStatement(sql);
        ps.setString(1, uname);
        ResultSet rs = ps.executeQuery();
        if(rs.next()){              //用户名存在
            b = true;
        }
        close(rs, ps, con);
        return b;
    }
    /**
     * 注册
     * @throws Exception
     */
    public boolean isRegist(String uname, String upass) throws Exception{
        boolean b = false;
        Connection con = getConnection();
        String sql = "insert into usertable values(?,?)";
        PreparedStatement ps = con.prepareStatement(sql);
        ps.setString(1, uname);
        ps.setString(2, upass);
        int i = ps.executeUpdate();
        if(i > 0){                  //注册成功
            b = true;
        }
        close(null, ps, con);
        return b;
    }
    /**
     * 登录
     * @throws Exception
     */
    public boolean isLogin(String uname, String upass) throws Exception{
        boolean b = false;          //用户名不存在
        Connection con = getConnection();
        String sql = "select * from usertable where userName = ? and password = ?";
        PreparedStatement ps = con.prepareStatement(sql);
        ps.setString(1, uname);
        ps.setString(2, upass);
```

```java
        ResultSet rs = ps.executeQuery();
        if(rs.next()){              //登录成功
            b = true;
        }
        close(rs, ps, con);
        return b;
    }
    /**
     * 修改密码
     * @throws Exception
     */
    public boolean updatePWD(String uname, String upass) throws Exception{
        boolean b = false;
        Connection con = getConnection();
        String sql = "update usertable set password = ? where userName = ?";
        PreparedStatement ps = con.prepareStatement(sql);
        ps.setString(1, upass);
        ps.setString(2, uname);
        int i = ps.executeUpdate();
        if(i > 0){                  //修改成功
            b = true;
        }
        close(null, ps, con);
        return b;
    }
}
```

## 10.5　名片管理

与系统相关的 JSP 页面、CSS 和图片位于 WebRoot 目录下。在 10.4 节中，已经介绍了系统的数据库操作，所以本节只是介绍 JSP 页面和 Action 的实现。

### 10.5.1　Action 的实现

CardAction 类负责处理"名片管理"的功能，包括添加、修改、删除、查询等，具体代码如下：

```java
package action;
import java.io.File;
import java.io.IOException;
import java.util.ArrayList;
import java.util.Map;
import org.apache.commons.io.FileUtils;
import org.apache.struts2.ServletActionContext;
import org.apache.struts2.interceptor.SessionAware;
import util.MyUtil;
import com.opensymphony.xwork2.ActionSupport;
import com.opensymphony.xwork2.ModelDriven;
import dao.CardDao;
```

```java
import entity.Card;
public class CardAction extends ActionSupport implements ModelDriven<Card>,SessionAware{
    private static final long serialVersionUID = 1L;
    Map<String, Object> session;
    CardDao cd = new CardDao();
    //封装名片信息
    private Card card = new Card();
    String act;                    //接收判断动作,查询,删除查询,修改查询
    String ids[];                  //接收多个id,删除多个
    private int pageCur;
    private int totalCount = 0;    //计算总共有多少个名片
    private int totalPage = 0;
    File logo;                     //图片文件对象
    String logoFileName;
    //查询名片信息
    private ArrayList<Card> allCards;
    //查询一个名片
    private Card acard;
    /**
     * 添加名片信息
     */
    public String add(){
        //选择logo文件
        if(logoFileName != null){
            //上传文件的保存位置"/logos",该位置是指tomcat的webapps\cardManage\logos
            String realpath = ServletActionContext.getServletContext().getRealPath("/logos");
            //声明文件目录logos
            File file = new File(realpath);
            //如果目录不存在,则创建该目录
            if(!file.exists()){
                file.mkdirs();
            }
            //上传到workspace下,开发时使用,部署后这块功能可注释
            File file1 = new File("D:/MyEclipse2014 workspace/cardManage/WebRoot/logos");
            //实现文件上传
            String fileType = logoFileName.substring(logoFileName.lastIndexOf('.'));
            String newFileName = MyUtil.getStringID() + fileType;
            card.setNewFileName(newFileName);
            try {
                FileUtils.copyFile(logo, new File(file, newFileName));
                //上传到workspace下,开发时使用,部署后这块功能可注释
                FileUtils.copyFile(logo, new File(file1, newFileName));
            } catch (IOException e) {
                e.printStackTrace();
            }
        }
        //保存到数据库
        card.setUserName((String)session.get("userName"));
        try {
            if(cd.add(card)){
                return "addSuccess";
```

```java
            }
            return "addFail";
        } catch (Exception e) {
            e.printStackTrace();
            return ERROR;
        }
    }
    /**
     * 修改
     */
    public String update(){
        //选择了logo文件
        if(logoFileName != null){
            //上传文件的保存位置"/logos",该位置是指tomcat的webapps\cardManage\logos
            String realpath = ServletActionContext.getServletContext().getRealPath("/logos");
            //声明文件目录logos
            File file = new File(realpath);
            //如果目录不存在,则创建该目录
            if(!file.exists()){
                file.mkdirs();
            }
            //上传到workspace下,开发时使用,部署后这块功能可注释
            File file1 = new File("D:/MyEclipse2014 workspace/cardManage/WebRoot/logos");
            //实现文件上传
            String fileType = logoFileName.substring(logoFileName.lastIndexOf('.'));
            String newFileName = MyUtil.getStringID() + fileType;
            card.setNewFileName(newFileName);
            try {
                FileUtils.copyFile(logo, new File(file, newFileName));
                //上传到workspace下,开发时使用,部署后这块功能可注释
                FileUtils.copyFile(logo, new File(file1, newFileName));
            } catch (IOException e) {
                e.printStackTrace();
            }
        }
        //删除两个目录下原先的图片
        if(card.getOldFileName().length() > 0){
            //上传文件的保存位置"/logos",该位置是指tomcat的webapps\cardManage\logos
            String realpath = ServletActionContext.getServletContext().getRealPath("/logos");
            //声明文件目录logos
            File file = new File(realpath);
            File file1 = new File("D:/MyEclipse2014 workspace/cardManage/WebRoot/logos");
            File f1 = new File(file,card.getOldFileName());
            File f2 = new File(file1,card.getOldFileName());
            f1.delete();
            f2.delete();
        }
        //修改名片信息
        try {
            cd.update(card);
            return "updateSuccess";
```

```java
        } catch (Exception e) {
            e.printStackTrace();
            return ERROR;
        }
    }
    /**
     * 查询所有名片
     */
    public String query(){
        try {
            ArrayList<Card> acl = cd.queryAll((String)session.get("userName"));
            int temp = acl.size();
            setTotalCount(temp);
            if (temp == 0) {
                totalPage = 0;  //总页数
            } else {
                //返回大于或者等于指定表达式的最小整数
                totalPage = (int) Math.ceil((double) temp / 10);
            }
            if ((pageCur - 1) * 10 >= temp) {
                pageCur = pageCur - 1;
            }
            if (pageCur == 0) {
                pageCur = 1;
            }
        } catch (Exception e) {
            // TODO Auto-generated catch block
            e.printStackTrace();
            return ERROR;
        }
        try {
            allCards = cd.queryByPage(pageCur, (String)session.get("userName"));
        } catch (Exception e) {
            e.printStackTrace();
            return ERROR;
        }
        if("deleteSelect".equals(act)){
            return "deleteSelect";
        }
        if("updateSelect".equals(act)){
            return "updateSelect";
        }
        return "querySuccess";
    }
    /**
     * 查询一个名片
     */
    public String selectA(){
        try {
            acard = cd.selectA(card.getId());
        } catch (Exception e) {
            // TODO Auto-generated catch block
            e.printStackTrace();
```

```java
            return ERROR;
        }
        if("updateAcard".equals(act)){
            return "updateAcard";
        }
        return "detailcard";
    }
    /**
     * 删除
     */
    public String delete(){
        try {
            if("link".equals(act)){
                cd.delete(card.getId());
            }
            if("button".equals(act)){
                cd.delete(ids);
            }
        } catch (Exception e) {
            // TODO Auto-generated catch block
            e.printStackTrace();
            return ERROR;
        }
        return "deleteSucess";
    }
    @Override
    public Card getModel() {
        return card;
    }
    @Override
    public void setSession(Map<String, Object> arg0) {
        session = arg0;
    }
    public File getLogo() {
        return logo;
    }
    public void setLogo(File logo) {
        this.logo = logo;
    }
    public String getLogoFileName() {
        return logoFileName;
    }
    public void setLogoFileName(String logoFileName) {
        this.logoFileName = logoFileName;
    }
    public ArrayList<Card> getAllCards() {
        return allCards;
    }
    public void setAllCards(ArrayList<Card> allCards) {
        this.allCards = allCards;
    }
    public Card getAcard() {
        return acard;
```

```java
        }
        public void setAcard(Card acard) {
            this.acard = acard;
        }
        public String getAct() {
            return act;
        }
        public void setAct(String act) {
            this.act = act;
        }
        public String[] getIds() {
            return ids;
        }
        public void setIds(String[] ids) {
            this.ids = ids;
        }
        public int getPageCur() {
            return pageCur;
        }
        public void setPageCur(int pageCur) {
            this.pageCur = pageCur;
        }
        public int getTotalCount() {
            return totalCount;
        }
        public void setTotalCount(int totalCount) {
            this.totalCount = totalCount;
        }
        public int getTotalPage() {
            return totalPage;
        }
        public void setTotalPage(int totalPage) {
            this.totalPage = totalPage;
        }
    }
```

### 10.5.2 添加名片

用户输入客户名片的姓名、电话、E-Mail、单位、职务、地址、logo 后，单击"提交"按钮实现添加。如果成功，则跳转到查询页面；如果失败，则回到添加页面。

addCard.jsp 页面实现添加名片信息的输入界面，如图 10.5 所示。

图 10.5 添加名片页面

addCard.jsp 的代码如下：

```jsp
<%@ page language="java" contentType="text/html; charset=UTF-8" pageEncoding="UTF-8"%>
<%@taglib prefix="s" uri="/struts-tags" %>
<%
```

```jsp
        String path = request.getContextPath();
        String basePath = request.getScheme() + "://"
                + request.getServerName() + ":" + request.getServerPort()
                + path + "/";
%>
<!DOCTYPE html PUBLIC "-//W3C//DTD HTML 4.01 Transitional//EN" "http://www.w3.org/TR/html4/loose.dtd">
<html>
<head>
<base href="<%=basePath%>">
<title>addCard.jsp</title>
<link href="css/common.css" type="text/css" rel="stylesheet">
</head>
<body>
    <s:form action="card/addCard.action" method="post" enctype="multipart/form-data">
        <table border=1 style="border-collapse: collapse">
            <caption>
                <font size=4 face=华文新魏>添加名片</font>
            </caption>
            <tr>
                <td>姓名<font color="red">*</font></td>
                <td><s:textfield name="name"/></td>
            </tr>
            <tr>
                <td>电话<font color="red">*</font></td>
                <td><s:textfield name="telephone"/></td>
            </tr>
            <tr>
                <td>E-Mail</td>
                <td><s:textfield name="email"/></td>
            </tr>
            <tr>
                <td>单位</td>
                <td><s:textfield name="company"/></td>
            </tr>
            <tr>
                <td>职务</td>
                <td><s:textfield name="post"/></td>
            </tr>
            <tr>
                <td>地址</td>
                <td><s:textfield name="address"/></td>
            </tr>
            <tr>
                <td>logo</td>
                <td><s:file name="logo"/></td>
            </tr>
            <tr>
                <td align="center"><s:submit value="提交"/></td>
                <td align="left"><s:reset value="重置"/></td>
            </tr>
```

```
            </table>
        </s:form>
        <s:fielderror/>
</body>
</html>
```

单击图10.5中的"提交"按钮,将添加请求通过"card/addCard.action"提交给Action处理。配置文件card.xml根据请求路径找到对应Action类CardAction(10.5.1节)的add方法处理添加功能。添加成功跳转到查询Action,添加失败回到添加页面。

### 10.5.3 查询名片

管理员登录成功后,进入名片管理系统的主页面,在主页面中初始显示查询页面queryCards.jsp,查询页面运行效果如图10.6所示。

| 名片ID | 名称 | 单位 | 详情 |
|---|---|---|---|
| 1 | 2121 | 2121 | 详情 |
| 2 | aaaaaaaaaa | dssd | 详情 |
| 3 | bbb | 4334 | 详情 |
| 21 | 21323 |  | 详情 |
| 22 | 323 |  | 详情 |
| 23 | 323 |  | 详情 |
| 24 | 221 |  | 详情 |
| 25 | wewe |  | 详情 |
| 26 | 323 |  | 详情 |
| 27 | 3223 |  | 详情 |

共11条记录  第1页  上一页  下一页

图10.6 查询页面

选择主页面中"名片管理"菜单的"查询名片"命令,打开查询页面queryCards.jsp。"查询名片"命令超链接的目标地址是个Action。该Action的请求路径为"card/queryCard.action",配置文件card.xml根据请求路径找到对应Action类的query方法处理查询功能。在该方法中,根据动作类型("修改查询""查询"以及"删除查询"),将查询结果转发到不同页面。

在queryCards.jsp页面中单击"详情"超链接,打开名片详细信息页面detail.jsp。"详情"超链接的目标地址是个Action。该Action的请求路径为"card/selectACard.action"。配置文件card.xml根据请求路径找到对应Action类的selectA方法处理查询一个名片的功能。将查询结果转发给详细信息页面detail.jsp。名片详细信息页面如图10.7所示。

名片详细信息

| ID | 1 |
|---|---|
| 姓名 | 2121 |
| 电话 | 2121 |
| E-mail | 2121 |
| 单位 | 2121 |
| 地址 | |
| logo |  |

图10.7 名片详情

**queryCards.jsp 的代码如下：**

```jsp
<%@ page language="java" contentType="text/html; charset=UTF-8" pageEncoding="UTF-8"%>
<%@taglib prefix="s" uri="/struts-tags" %>
<%
String path = request.getContextPath();
String basePath = request.getScheme()+"://"+request.getServerName()+":"+request.getServerPort()+path+"/";
%>
<!DOCTYPE html PUBLIC "-//W3C//DTD HTML 4.01 Transitional//EN" "http://www.w3.org/TR/html4/loose.dtd">
<html>
  <head>
    <base href="<%=basePath%>">
    <title>queryCards.jsp</title>
    <link href="css/common.css" type="text/css" rel="stylesheet">
    <style type="text/css">
        table{
            text-align: center;
            border-collapse: collapse;
        }
        .bgcolor{
            background-color: #F08080;
        }
    </style>
    <script type="text/javascript">
        function changeColor(obj){
            obj.className = "bgcolor";
        }
        function changeColor1(obj){
            obj.className = "";
        }
    </script>
  </head>
  <body>
    <table border="1" bordercolor="PaleGreen">
        <tr>
            <th width="200px">名片ID</th>
            <th width="200px">名称</th>
            <th width="250px">单位</th>
            <th width="200px">详情</th>
        </tr>
        <!-- allCards 为action中的变量 -->
        <s:iterator value="allCards" var="card">
            <tr onmousemove="changeColor(this)" onmouseout="changeColor1(this)">
                <td><s:property value="id"/></td>
                <td><s:property value="name"/></td>
                <td><s:property value="company"/></td>
    <td><a href="card/selectACard.action?id=<s:property value="id"/>" target="_blank">详情</a></td>
```

```
            </tr>
        </s:iterator>
        <tr>
            <td colspan="4" align="right">
               
               共${totalCount}条记录   
            第${pageCur}页   
            <s:url id="url_pre" value="card/queryCard.action">
                <s:param name="pageCur" value="pageCur-1"></s:param>
            </s:url>
            <s:url id="url_next" value="card/queryCard.action">
                <s:param name="pageCur" value="pageCur+1"></s:param>
            </s:url>
            <!-- OGNL 语言默认会调用-1 的方法 -->
            <s:a href="%{url_pre}">上一页</s:a>
                
            <s:a href="%{url_next}">下一页</s:a>
            </td>
        </tr>
    </table>
</body>
</html>
```

detail.jsp 的代码如下：

```
<%@ page language="java" contentType="text/html; charset=UTF-8" pageEncoding="UTF-8"%>
<%@taglib prefix="s" uri="/struts-tags" %>
<%
    String path = request.getContextPath();
    String basePath = request.getScheme() + "://" + request.getServerName() + ":" + request.getServerPort() + path + "/";
%>
<!DOCTYPE html PUBLIC "-//W3C//DTD HTML 4.01 Transitional//EN" "http://www.w3.org/TR/html4/loose.dtd">
<html>
  <head>
    <title>detail.jsp</title>
    <base href="<%=basePath%>">
  </head>
  <body>
    <center>
        <table border=1 background="images/bb.jpg" style="border-collapse: collapse">
            <caption>
                <font size=4 face=华文新魏>名片详细信息</font>
            </caption>
            <tr>
                <td>ID</td>
                <td><s:property value="acard.id"/></td>
            </tr>
            <tr>
                <td>姓名</td>
```

```html
                    <td><s:property value = "acard.name"/></td>
                </tr>
                <tr>
                    <td>电话</td>
                    <td><s:property value = "acard.telephone"/></td>
                </tr>
                <tr>
                    <td>E-mail</td>
                    <td><s:property value = "acard.email"/></td>
                </tr>
                <tr>
                    <td>单位</td>
                    <td><s:property value = "acard.company"/></td>
                </tr>
                <tr>
                    <td>地址</td>
                    <td><s:property value = "acard.address"/></td>
                </tr>
                <tr>
                    <td>logo</td>
                    <td>
                        <s:if test = "acard.newFileName != null">
                            <img alt = "" width = "250" height = "250"
                            src = "logos/<s:property value = "acard.newFileName"/>"/>
                        </s:if>
                        <s:else>
                            没有logo
                        </s:else>
                    </td>
                </tr>
            </table>
        </center>
    </body>
</html>
```

## 10.5.4 修改名片

选择主页面中"管理名片"菜单的"修改名片"命令,打开修改查询页面 updateSelect.jsp。"修改名片"命令超链接的目标地址是个 Action。找到对应 Action 类 CardAction 的方法 query,在该方法中,根据动作类型,将查询结果转发给修改查询页面。

单击 updateSelect.jsp 页面中的"修改"超链接打开修改名片信息页面 updateCard.jsp。"修改"超链接的目标地址是个 Action。找到对应 Action 类 CardAction 的方法 selectA,在该方法中,根据动作类型,将查询结果转发给 updateCard.jsp 页面显示。

输入要修改的信息后,单击"提交"按钮,将名片信息提交给 Action,找到对应 Action 类 CardAction 的方法 update,在方法中执行修改的业务处理。修改成功,进入查询名片。修改失败,回到 updateCard.jsp 页面。

updateSelect.jsp 页面的运行效果如图 10.8 所示,updateCard.jsp 页面的运行效果如图 10.9 所示。

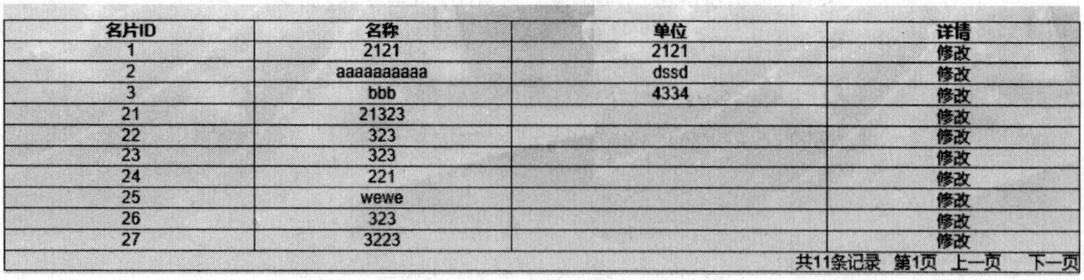

| 名片ID | 名称 | 单位 | 详情 |
|---|---|---|---|
| 1 | 2121 | 2121 | 修改 |
| 2 | aaaaaaaaaa | dssd | 修改 |
| 3 | bbb | 4334 | 修改 |
| 21 | 21323 | | 修改 |
| 22 | 323 | | 修改 |
| 23 | 323 | | 修改 |
| 24 | 221 | | 修改 |
| 25 | wewe | | 修改 |
| 26 | 323 | | 修改 |
| 27 | 3223 | | 修改 |

共11条记录　第1页　上一页　下一页

图 10.8　updateSelect.jsp 页面

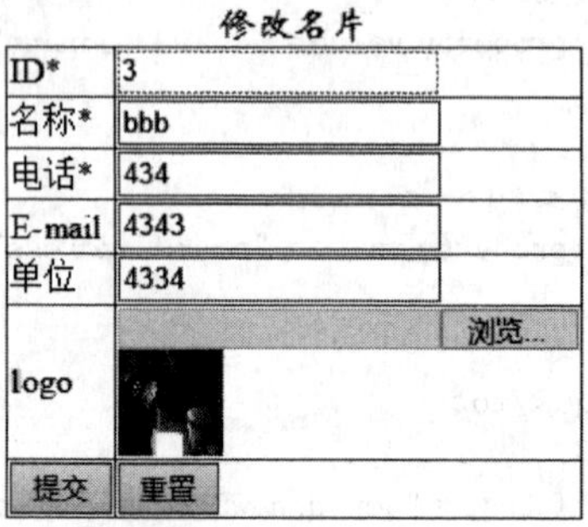

图 10.9　updateCard.jsp 页面

updateSelect.jsp 的代码如下：

```jsp
<%@ page language = "java" contentType = "text/html; charset = UTF-8" pageEncoding = "UTF-8" %>
<%@taglib prefix = "s" uri = "/struts-tags" %>
<%
    String path = request.getContextPath();
    String basePath = request.getScheme() + "://" + request.getServerName() + ":" + request.getServerPort() + path + "/";
%>
<!DOCTYPE html PUBLIC "-//W3C//DTD HTML 4.01 Transitional//EN" "http://www.w3.org/TR/html4/loose.dtd">
<html>
  <head>
    <base href = "<% = basePath %>">
    <title>updateSelect.jsp</title>
    <link href = "css/common.css" type = "text/css" rel = "stylesheet">
    <style type = "text/css">
        table{
            text-align: center;
            border-collapse: collapse;
        }
        .bgcolor{
            background-color: #F08080;
        }
    </style>
    <script type = "text/javascript">
        function changeColor(obj){
```

```html
                obj.className = "bgcolor";
            }
            function changeColor1(obj){
                obj.className = "";
            }
        </script>
    </head>
    <body>
        <br>
        <table border="1" bordercolor="PaleGreen">
            <tr>
                <th width="200px">名片ID</th>
                <th width="200px">名称</th>
                <th width="250px">单位</th>
                <th width="200px">详情</th>
            </tr>
            <!-- allCards 为 action 中的变量 -->
            <s:iterator value="allCards" var="card">
            <tr onmousemove="changeColor(this)" onmouseout="changeColor1(this)">
                <td><s:property value="id"/></td>
                <td><s:property value="name"/></td>
                <td><s:property value="company"/></td>
                <td><a href="card/selectACard.action?id=<s:property value="id"/>&act=updateAcard" target="center">修改</a></td>
            </tr>
            </s:iterator>
            <tr>
                <td colspan="4" align="right">
                   
                   共${totalCount}条记录  
                第${pageCur}页  
                <s:url id="url_pre" value="card/queryCard.action">
                    <s:param name="pageCur" value="pageCur-1"></s:param>
                    <s:param name="act" value="'updateSelect'"></s:param>
                </s:url>
                <s:url id="url_next" value="card/queryCard.action">
                    <s:param name="pageCur" value="pageCur+1"></s:param>
                    <s:param name="act" value="'updateSelect'"></s:param>
                </s:url>
                <!-- OGNL 语言默认会调用-1的方法 -->
                <s:a href="%{url_pre}">上一页</s:a>
                    
                <s:a href="%{url_next}">下一页</s:a>
                </td>
            </tr>
        </table>
    </body>
</html>
```

updateCard.jsp 的代码如下：

```jsp
<%@ page language="java" import="java.util.*" pageEncoding="UTF-8"%>
<%@taglib prefix="s" uri="/struts-tags" %>
<%
    String path = request.getContextPath();
    String basePath = request.getScheme()+"://"+request.getServerName()+":"+request.getServerPort()+path+"/";
%>
<!DOCTYPE HTML PUBLIC "-//W3C//DTD HTML 4.01 Transitional//EN">
<html>
  <head>
    <base href="<%=basePath%>">
    <title>My JSP 'updateCard.jsp' starting page</title>
  </head>
  <body>
    <s:form action="card/updateCard.action" method="post" enctype="multipart/form-data">
        <table border=1 style="border-collapse:collapse">
            <caption>
                <font size=4 face=华文新魏>修改名片</font>
            </caption>
            <tr>
                <td>ID<font color="red">*</font></td>
                <td><s:textfield name="id"
                cssStyle="border-width:1pt;border-style:dashed;border-color:red"
                 value="%{acard.id}"
                readonly="true"/>
                </td>
            </tr>
            <tr>
                <td>名称<font color="red">*</font></td>
                <td><s:textfield name="name" value="%{acard.name}"/></td>
            </tr>
            <tr>
                <td>电话<font color="red">*</font></td>
                <td><s:textfield name="telephone" value="%{acard.telephone}"/></td>
            </tr>
            <tr>
                <td>E-mail</td>
                <td><s:textfield name="email" value="%{acard.email}"/></td>
            </tr>
            <tr>
                <td>单位</td>
                <td><s:textfield name="company" value="%{acard.company}"/></td>
            </tr>
            <tr>
                <td>logo</td>
                <td>
                    <s:file name="logo"/><br>
                    <s:if test="acard.newFileName != null">
                        <img alt="" width="50" height="50"
                        src="logos/<s:property value="acard.newFileName"/>"/>
                    </s:if>
```

```
                <s:hidden name = "oldFileName" value = " % {acard.newFileName}"/>
            </td>
        </tr>
        <tr>
            <td align = "center"><s:submit value = "提交"/></td>
            <td align = "left"><s:reset value = "重置"/></td>
        </tr>
    </table>
  </s:form>
</body>
</html>
```

### 10.5.5 删除名片

选择主页面中"管理名片"菜单的"删除名片"命令，打开删除查询页面 deleteSelect.jsp。"删除名片"命令超链接的目标地址是个 Action。找到对应 Action 类 CardAction 的方法 query，在该方法中，根据动作类型，将查询结果转发给 deleteSelect.jsp 页面，页面效果如图 10.10 所示。

| ID | 名称 | 单位 | 详情 | 操作 |
|---|---|---|---|---|
| ☐1 | 2121 | 2121 | 详情 | 删除 |
| ☐2 | aaaaaaaaaa | dssd | 详情 | 删除 |
| ☐3 | bbb | 4334 | 详情 | 删除 |
| ☐21 | 21323 | | 详情 | 删除 |
| ☐22 | 323 | | 详情 | 删除 |
| ☐23 | 323 | | 详情 | 删除 |
| ☐24 | 221 | | 详情 | 删除 |
| ☐25 | wewe | | 详情 | 删除 |
| ☐26 | 323 | | 详情 | 删除 |
| ☐27 | 3223 | | 详情 | 删除 |
| | | 删除 | | |
| | | | 共11条记录　第1页　上一页 | 下一页 |

图 10.10　deleteSelect.jsp 页面

在图 10.10 的复选框中选中要删除的名片，单击"删除"按钮，将要删除名片的 ID 提交给控制器 Action。找到对应 Action 类 CardAction 的方法 delete，在该方法中，根据动作类型执行批量删除的业务处理。

单击图 10.10 中的"删除"超链接，将当前行的名片 ID 提交给控制器 Action，找到对应 Action 类 CardAction 的方法 delete，在该方法中，根据动作类型执行单个删除的业务处理。

删除成功后，进入删除查询页面。

deleteSelect.jsp 的代码如下：

```
<%@ page language = "java" contentType = "text/html; charset = UTF - 8" pageEncoding =
"UTF - 8" %>
<%@taglib prefix = "s" uri = "/struts - tags" %>
<%
    String path = request.getContextPath();
    String basePath = request.getScheme() + "://" + request.getServerName() + ":" + request.
    getServerPort() + path + "/";
%>
<! DOCTYPE html PUBLIC " - //W3C//DTD HTML 4.01 Transitional//EN" "http://www.w3.org/TR/html4/
```

```
loose.dtd">
<html>
  <head>
    <base href="<%=basePath%>">
    <title>deleteSelect.jsp</title>
    <link href="css/common.css" type="text/css" rel="stylesheet">
    <style type="text/css">
        table{
            text-align: center;
            border-collapse: collapse;
        }
        .bgcolor{
            background-color: #F08080;
        }
    </style>
    <script type="text/javascript">
        function confirmDelete(){
            var n = document.deleteForm.ids.length;
            var count = 0;     //统计没有选中的个数
            for(var i = 0; i < n; i++){
                if(!document.deleteForm.ids[i].checked){
                    count++;
                }else{
                    break;
                }
            }
            if(n > 1){         //多个名片
                //所有的名片都没有选择
                if(count == n){
                    alert("请选择删除的名片!");
                    count = 0;
                    return false;
                }
            }else{             //一个名片
                //就一个名片并且还没有选择
                if(!document.deleteForm.ids.checked){
                    alert("请选择删除的名片!");
                    return false;
                }
            }
            if(window.confirm("真的删除吗?really?")){
                document.deleteForm.submit();
                return true;
            }
            return false;
        }
        function checkDel(id){
            if(window.confirm("是否删除该名片?")){
                window.location.href = "/cardManage/card/deleteCard.action?act=link&id=" + id;
            }
```

```
            }
            function changeColor(obj){
                obj.className = "bgcolor";
            }
            function changeColor1(obj){
                obj.className = "";
            }
    </script>
</head>
<body>
    <br>
    <s:form action = "card/deleteCard.action?act = button" name = "deleteForm">
    <table border = "1" bordercolor = "PaleGreen">
        <tr>
            <th width = "250px">ID</th>
            <th width = "200px">名称</th>
            <th width = "200px">单位</th>
            <th width = "200px">详情</th>
            <th width = "200px">操作</th>
        </tr>
        <!-- allCards 为 action 中的变量 -->
        <s:iterator value = "allCards" var = "c">
            <tr onmousemove = "changeColor(this)" onmouseout = "changeColor1(this)">
                <td>
                    <input type = "checkbox" name = "ids" value = "<s:property value = "id"/>"/>
                    <s:property value = "id"/>
                </td>
                <td><s:property value = "name"/></td>
                <td><s:property value = "company"/></td>
                <td><a href = "card/selectACard.action?id = <s:property value = "id"/>" target = "_blank">详情</a></td>
                <td>
                    <a href = "javascript:checkDel('<s:property value = "id"/>')">删除</a>
                </td>
            </tr>
        </s:iterator>
        <tr>
            <td colspan = "5">
                <input type = "button" value = "删除" onclick = "confirmDelete()">
            </td>
        </tr>
        <tr>
            <td colspan = "5" align = "right">
                   
                   共 ${totalCount}条记录   
                第 ${pageCur}页   
                <s:url id = "url_pre" value = "card/queryCard.action">
                    <s:param name = "pageCur" value = "pageCur - 1"></s:param>
                    <s:param name = "act" value = "'deleteSelect'"></s:param>
                </s:url>
```

```
            <s:url id="url_next" value="card/queryCard.action">
                <s:param name="pageCur" value="pageCur+1"></s:param>
                <s:param name="act" value="'deleteSelect'"></s:param>
            </s:url>
            <!-- OGNL语言默认会调用-1的方法 -->
            <s:a href="%{url_pre}">上一页</s:a>
                
            <s:a href="%{url_next}">下一页</s:a>
            </td>
        </tr>
    </table>
    </s:form>
  </body>
</html>
```

## 10.6 用户相关

### 10.6.1 Action的实现

UserAction类负责处理"会员注册""会员登录""安全退出"以及"个人中心"的功能,具体代码如下:

```
package action;
import java.util.Map;
import org.apache.struts2.interceptor.RequestAware;
import org.apache.struts2.interceptor.SessionAware;
import com.opensymphony.xwork2.ActionSupport;
import dao.UserDao;
public class UserAction extends ActionSupport implements RequestAware,SessionAware{
    private static final long serialVersionUID = 1L;
    //标记位,为0 判断用户名是否可用;为1 注册
    private String flag;
    private String uname;
    private String upass;
    private Map<String, Object> request;
    private Map<String, Object> session;
    UserDao ud = new UserDao();
    /**
     * 注册
     */
    public String register(){
        try {
            if("0".equals(flag)){           //判断用户名是否可用
                if(ud.isExit(uname)){       //用户名存在
                    request.put("isExit", "false");
                }else{
                    request.put("isExit", "true");
                }
```

```java
                request.put("uname", uname);
                return "register";
            }else{                              //实现注册功能
                if(ud.isRegist(uname, upass)){
                    return SUCCESS;
                }else{
                    return "register";
                }
            }
        } catch (Exception e) {
            // TODO Auto-generated catch block
            e.printStackTrace();
            return ERROR;
        }
    }
    /**
     * 登录
     */
    public String login(){
        try {
            if(ud.isLogin(uname, upass)){
                session.put("userName", uname);
                session.put("userPWD", upass);
                return SUCCESS;
            }
            //在页面使用<s:fielderror/>取出错误信息
            this.addFieldError("fail", "用户名或密码错误!");
            return "loginFail";
        } catch (Exception e) {
            // TODO Auto-generated catch block
            e.printStackTrace();
            return ERROR;
        }
    }
    /**
     * 修改密码
     */
    public String updatePwd(){
        try {
            ud.updatePWD((String)session.get("userName"), upass);
        } catch (Exception e) {
            e.printStackTrace();
            return ERROR;
        }
        return SUCCESS;
    }
    /**
     * 安全退出
     */
    public String logout(){
        session.clear();
```

```
            return SUCCESS;
        }
        public String getFlag() {
            return flag;
        }
        public void setFlag(String flag) {
            this.flag = flag;
        }
        public String getUname() {
            return uname;
        }
        public void setUname(String uname) {
            this.uname = uname;
        }
        public String getUpass() {
            return upass;
        }
        public void setUpass(String upass) {
            this.upass = upass;
        }
        @Override
        public void setRequest(Map<String, Object> arg0) {
            request = arg0;
        }
        @Override
        public void setSession(Map<String, Object> arg0) {
            session = arg0;
        }
    }
```

## 10.6.2 注册

在系统默认主页 index.jsp,单击"注册"链接,打开注册页面 register.jsp,效果如图 10.11 所示。

在图 10.11 所示的注册页面中,输入"姓名"后,系统会根据请求路径"user/regist.action"和标记位"flag"检测"姓名"是否可用。输入合法的用户信息后,单击"注册"按钮,实现注册功能。

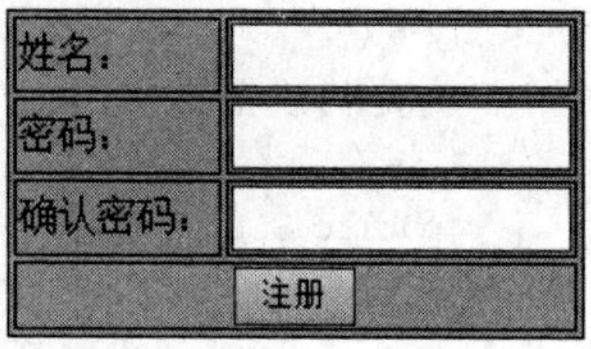

register.jsp 的代码如下:

图 10.11 注册页面

```
<%@ page language="java" contentType="text/html; charset=GBK" pageEncoding="GBK" %>
<%@ taglib prefix="s" uri="/struts-tags" %>
<%
    String path = request.getContextPath();
    String basePath = request.getScheme() + "://" + request.getServerName() + ":" + request.getServerPort() + path + "/";
%>
<!DOCTYPE html PUBLIC "-//W3C//DTD HTML 4.01 Transitional//EN" "http://www.w3.org/TR/html4/
```

```html
loose.dtd">
<html>
    <head>
    <base href="<%=basePath%>">
    <style type="text/css">
        .textSize{
            width:100pt;
            height:15pt
        }
    </style>
    <title>注册画面</title>
    <script type="text/javascript">
        //输入姓名后,调用该方法,判断用户名是否可用
        function nameIsNull(){
            var name = document.registForm.uname.value;
            if(name == ""){
                alert("请输入姓名!");
                document.registForm.uname.focus();
                return false;
            }
            document.registForm.flag.value = "0";
            document.registForm.submit();
            return true;
        }
        //注册时检查输入项
        function allIsNull(){
            var name = document.registForm.uname.value;
            var pwd = document.registForm.upass.value;
            var repwd = document.registForm.reupass.value;
            if(name == ""){
                alert("请输入姓名!");
                document.registForm.uname.focus();
                return false;
            }
            if(pwd == ""){
                alert("请输入密码!");
                document.registForm.upass.focus();
                return false;
            }
            if(repwd == ""){
                alert("请输入确认密码!");
                document.registForm.reupass.focus();
                return false;
            }
            if(pwd != repwd){
                alert("2次密码不一致,请重新输入!");
                document.registForm.upass.value = "";
                document.registForm.reupass.value = "";
```

```html
                document.registForm.upass.focus();
                return false;
            }
            document.registForm.flag.value="1";
            document.registForm.submit();
            return true;
        }
    </script>
</head>
<body>
    <form action="user/regist.action" method="post" name="registForm">
        <input type="hidden" name="flag">
        <table border=1 bgcolor="lightblue" align="center">
            <tr>
                <td>姓名:</td>
                <td>
                    <input class="textSize" type="text" name="uname" value="${requestScope.uname}" onblur="nameIsNull()" />
                    <s:if test="#request.isExit=='false'">
                        <font color=red size=5>×</font>
                    </s:if>
                    <s:if test="#request.isExit=='true'">
                        <font color=green size=5>√</font>
                    </s:if>
                </td>
            </tr>
            <tr>
                <td>密码:</td>
                <td><input class="textSize" type="password" maxlength="20" name="upass"/></td>
            </tr>
            <tr>
                <td>确认密码:</td>
                <td><input class="textSize" type="password" maxlength="20" name="reupass"/></td>
            </tr>
            <tr>
                <td colspan="2" align="center"><input type="button" value="注册" onclick="allIsNull()"/></td>
            </tr>
        </table>
    </form>
</body>
</html>
```

### 10.6.3 登录

在系统默认主页 index.jsp,单击"登录"链接,打开登录页面 login.jsp,效果如图 10.12 所示。

用户输入姓名和密码后,系统将对姓名和密码进行验证。如果姓名和密码同时正确,则成功登录,将用户信息保存到 session 对象,并进入系统管理主页面(main.jsp);如果姓名或密码有误,则提示错误。

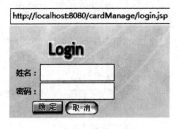

图 10.12 登录界面

login.jsp 的代码如下:

```
<%@ page language="java" contentType="text/html; charset=UTF-8"
    pageEncoding="UTF-8"%>
<%@taglib prefix="s" uri="/struts-tags" %>
<%
    String path = request.getContextPath();
    //获取当前项目的路径,如: http://localhost:8080/项目名称.
    String basePath = request.getScheme() + "://" + request.getServerName() + ":" + request.getServerPort() + path + "/";
%>
<!DOCTYPE html PUBLIC "-//W3C//DTD HTML 4.01 Transitional//EN" "http://www.w3.org/TR/html4/loose.dtd">
<html>
  <head>
    <base href="<%=basePath%>">
    <!--
    base 标记是一个基链接标记,是一个单标记.用以改变文件中所有链接标记的参数内定值,网页上的所有相对路径在链接时都将在前面加上基链接指向的地址.
    比如:<base href="http://www.baidu.com">,那下面的 href 属性就会以上面设的为基准,方便使用相对路径
    如:<a href="http://www.baidu.com/xxx.htm"></a>,现在就只需要写<a href="xxx.htm"></a>
    -->
    <title>后台登录</title>
    <style type="text/css">
    table{
        text-align: center;
    }
    .textSize{
        width: 120px;
        height: 25px;
    }
    * {
        margin: 0px;
        padding: 0px;
    }
    body {
        font-family: Arial, Helvetica, sans-serif;
```

```
            font-size: 12px;
            margin: 10px 10px auto;
            background-image: url(images/bb.jpg);
        }
        .errorMessage {font-weight:bold; color:red; }
        <!-- errorMessage 会自动更改错误消息的样式 -->
        </style>
        <script type = "text/javascript">
        //确定按钮
        function gogo(){
            document.forms[0].submit();
        }
        //取消按钮
        function cancel(){
            document.forms[0].action = "";
        }
        </script>
    </head>
    <body>
        <s:form action = "user/login.action" method = "post">
        <table>
            <tr>
                <td colspan = "2"><img src = "images/login.gif"></td>
            </tr>
            <tr>
                <td>姓名：</td>
                <td><s:textfield name = "uname" cssClass = "textSize"/></td>
            </tr>
            <tr>
                <td>密码：</td>
                <td><s:password name = "upass" cssClass = "textSize"/></td>
            </tr>
            <tr>
                <td colspan = "2">
                    <input type = "image" src = "images/ok.gif" onclick = "gogo()">
                    <input type = "image" src = "images/cancel.gif" onclick = "cancel()">
                </td>
            </tr>
        </table>
        </s:form>
        <s:fielderror/>
    </body>
</html>
```

单击图10.12中的"确定"按钮，通过请求路径"user/login.action"，将登录请求提交给Action。配置文件user.xml根据请求路径找到对应Action类UserAction(10.6.1节)的login方法处理登录请求。

### 10.6.4 修改密码

选择主页面中"个人中心"菜单的"修改密码"命令,打开密码修改页面 updatePWD.jsp,页面效果如图 10.13 所示。

图 10.13 密码修改页面

updatePWD.jsp 的代码如下:

```jsp
<%@ page language="java" import="java.util.*" pageEncoding="UTF-8"%>
<%@taglib prefix="s" uri="/struts-tags" %>
<%
    String path = request.getContextPath();
    String basePath = request.getScheme()+"://"+request.getServerName()+":"+request.getServerPort()+path+"/";
%>
<!DOCTYPE html PUBLIC "-//W3C//DTD HTML 4.01 Transitional//EN" "http://www.w3.org/TR/html4/loose.dtd">
<html>
  <head>
    <base href="<%=basePath%>">
    <style type="text/css">
        table{
            text-align: center;
        }
        .textSize{
            width: 120px;
            height: 25px;
        }
        * {
            margin: 0px;
            padding: 0px;
        }
        body {
            font-family: Arial, Helvetica, sans-serif;
            font-size: 12px;
            margin: 10px 10px auto;
            background-image: url(images/bb.jpg);
        }
    </style>
    <title>修改密码</title>
    <script type="text/javascript">
        //注册时检查输入项
        function allIsNull(){
```

```html
                var pwd = document.updateForm.upass.value;
                var repwd = document.updateForm.reupass.value;
                if(pwd == ""){
                    alert("请输入新密码!");
                    document.updateForm.upass.focus();
                    return false;
                }
                if(repwd == ""){
                    alert("请输入确认新密码!");
                    document.updateForm.reupass.focus();
                    return false;
                }
                if(pwd != repwd){
                    alert("2次密码不一致,请重新输入!");
                    document.updateForm.upass.value = "";
                    document.updateForm.reupass.value = "";
                    document.updateForm.upass.focus();
                    return false;
                }
                document.updateForm.submit();
                return true;
            }
        </script>
    </head>
    <body>
        <form action = "user/updatePwd.action" method = "post" name = "updateForm">
            <table>
                <tr>
                    <td>姓名:</td>
                    <td>
                        <s:property value = "#session.userName"/>
                    </td>
                </tr>
                <tr>
                    <td>新密码:</td>
                    <td><input class = "textSize" type = "password" maxlength = "20" name = "upass"/></td>
                </tr>
                <tr>
                    <td>确认新密码:</td>
                    <td><input class = "textSize" type = "password" maxlength = "20" name = "reupass"/></td>
                </tr>
                <tr>
                    <td colspan = "2" align = "center"><input type = "button" value = "修改密码" onclick = "allIsNull()"/></td>
                </tr>
            </table>
        </form>
    </body>
</html>
```

在图 10.13 中输入"新密码"和"确认新密码"后,单击"修改密码"按钮,将请求通过"user/updatePwd.action"提交给 Action。配置文件 user.xml 根据请求路径找到对应 Action 类 UserAction(10.6.1 节)的 updatePwd 方法处理密码修改请求。

### 10.6.5 基本信息

选择主页面中"个人中心"菜单的"基本信息"命令,打开基本信息页面 userInfo.jsp。页面效果如图 10.14 所示。

图 10.14 基本信息页面

userInfo.jsp 的代码如下:

```jsp
<%@ page language="java" contentType="text/html; charset=UTF-8"
    pageEncoding="UTF-8"%>
<%@taglib prefix="s" uri="/struts-tags" %>
<%
    String path = request.getContextPath();
    String basePath = request.getScheme() + "://" + request.getServerName() + ":" + request.getServerPort() + path + "/";
%>
<!DOCTYPE html PUBLIC "-//W3C//DTD HTML 4.01 Transitional//EN" "http://www.w3.org/TR/html4/loose.dtd">
<html>
  <head>
    <base href="<%=basePath%>">
    <title>用户基本信息</title>
    <style type="text/css">
    table{
        text-align: center;
    }
    .textSize{
        width: 120px;
        height: 25px;
    }
    * {
        margin: 0px;
        padding: 0px;
    }
    body {
        font-family: Arial, Helvetica, sans-serif;
        font-size: 12px;
        margin: 10px 10px auto;
        background-image: url(images/bb.jpg);
    }
    </style>
  </head>
  <body>
    <table>
        <tr>
            <td colspan="2">用户基本信息</td>
        </tr>
        <tr>
```

```
            <td>姓名:</td>
            <td><s:property value="#session.userName"/></td>
        </tr>
        <tr>
            <td>密码:</td>
            <td><input type="password" readonly value="<s:property value="#session.userPWD"/>"></td>
        </tr>
    </table>
  </body>
</html>
```

## 10.7 安全退出

在管理主页面中,单击"安全退出"超链接,将返回后台登录页面。"安全退出"超链接的目标地址是一个 Action,找到对应 Action 类 UserAction 的方法 logout。在该方法中执行:

session.clear();

将登录信息清除,并返回登录页面。

## 10.8 本章小结

本章讲述了名片管理系统的设计与实现。通过本章的学习,读者不仅掌握 Struts 2 应用开发的流程、方法和技术,还应该熟悉名片管理的业务需求、设计以及实现。

# 参 考 文 献

[1] 陈恒,张一鸣.Struts 2 框架应用教程[M].北京:清华大学出版社,2016.
[2] 史胜辉,王春明,陆培军.JavaEE 轻量级框架 Struts 2+Spring+Hibernate 整合开发[M].北京:清华大学出版社,2014.
[3] 孙连伟,武迪.Struts 2 程序开发实用教程[M].北京:清华大学出版社,2014.
[4] 王永贵,郭伟,冯永安,等.Java 高级框架应用开发案例教程——Struts 2+Spring+Hibernate [M].北京:清华大学出版社,2012.
[5] 范新灿.基于 Struts、Hibernate、Spring 架构的 Web 应用开发[M].2 版.北京:电子工业出版社,2014.
[6] 李刚.Struts 2 权威指南——基于 WebWork 核心的 MVC 开发[M].北京:电子工业出版社,2007.